# ROGER SHERMAN

*and the Independent Oil Men*

# ROGER SHERMAN

## *and the Independent Oil Men*

---

CHESTER McARTHUR DESTLER

*Cornell University Press*

ITHACA, NEW YORK

CORNELL UNIVERSITY PRESS

*First published 1967*

Library of Congress Catalog Card Number: 67–13466

PRINTED IN THE UNITED STATES OF AMERICA
BY KINGSPORT PRESS, INC.

# *Preface and Acknowledgments*

ROGER SHERMAN of Titusville, Pennsylvania, was a liberal. Inheriting the Jacksonian tradition from cultivated New York parents, he grew up in the antebellum southwest, where he began to practice law on the eve of the Civil War. Almost two years' service with General Nathan B. Forrest's cavalry oriented him toward the Confederate tradition. These attitudes he combined with attachment to high ethical and cultural standards. Ruined by the war, parentless but sponsored by his Conkling relatives of upstate New York, he went north after leaving Forrest's corps, and returned to the law. He settled first in Pithole in the Oil Region. Ruined again by the great fire at Pithole, he re-established his practice, moved eventually to Titusville, and rose steadily to a position of influence. He became a specialist in equity jurisprudence.

As a leader of the bar in northwestern Pennsylvania he became the chief legal adviser and strategist of the independent oil men,* with whom he worked to challenge the rising monopoly of the Standard Oil Company. The champion of competition and free enterprise in the petroleum industry, he contributed notably to the independents' recurring campaigns against the Rockefeller coterie. At the same time, in his other activities, he worked steadily for the reform of business ethics; he contributed significantly to the development of the regulatory state and fought the "Boss" system of politics and the great interests that employed it in Pennsylvania. While he helped to fashion a new legislative approach to the business problems of the era, his main efforts were directed toward the ap-

* The phrase "the independent oil men" was used during the last three decades of the nineteenth century to describe Sherman and his cohorts.

peal to the courts to impose effective restraints upon the Rockefellers and their railroad allies during "The Gilded Age."

Interestingly, although he joined the Civil Service Reform League, he also emerged as an influential anti-"Ring," antimonopolist leader of the Democratic Party. Because of the pertinence of his leadership in the petroleum industry to the larger antimonopoly campaign, Sherman became a figure of national importance. This enabled him to exert significant influence upon the movements for national railroad regulation and for restraint of industrial monopolies. He became a friend and consultant of Henry Demarest Lloyd, America's leading antimonopolist, and helped Lloyd to establish the factual accuracy and adequacy of *Wealth Against Commonwealth.* Prodded by Lloyd, Sherman successfully urged upon the independent oil men a policy of integration and the invoking of the criminal provisions of the Sherman Act against the Standard Oil executives.

Meanwhile, in Titusville he became a civic leader who contributed significantly to urban rebuilding, to cultural activities, and to journalism. In many respects he was a prototype of the later Progressives in his thinking, as evidenced in his reform activity, his political leadership, and in his insistence upon appeal to the courts and the legislatures to remedy problems produced by the growth of special privilege and the rise of "Big Business."

Preparation of this study was undertaken as an independent project after the discovery of the Roger Sherman Papers at the residence of T. W. Phillips, Jr., of Butler, Pennsylvania. Mr. Phillips extended unusual courtesies to me there during my first work on the papers, and then gave them and Roger Sherman's private library to Yale University Library, where Librarian James T. Babb and the Curators of Historical Manuscripts, Mrs. Zara Powers and Mr. Howard B. Gotlieb, facilitated my research.

Curators Paul H. Giddens and James G. Kehem of the Drake Well Museum and their staff extended the usual courtesies during my research there. Miss Alice E. Smith, Curator of Manuscripts, facilitated my research in the Henry Demarest Lloyd Papers at the State Historical Society of Wisconsin. The late William Bross Lloyd made available the Henry Demarest Lloyd Papers in Winnetka, Illinois, which have since been contributed to the State Historical Society of Wisconsin. Mr. James B. Stevenson, editor of the *Titusville Morning Herald,* lent me the important manuscript of Mrs. W. B. Roberts' Diary and extended to me every courtesy during my work on the long file of his paper in his office. Mrs. Hilda E. Cotton, Librarian, and the staff of the Benson Memorial Library of Titusville were especially helpful. In addition to these important sources of information I enjoyed the courtesy of interviews with Mrs. Lena Emery Brenneman, a daughter of Lewis Emery, Jr., and with Miss Lucy Grumbine, the daughter of Roger Sherman's last law partner.

CHESTER MCARTHUR DESTLER

*West Hartford, Connecticut*
*September 1966*

# *Contents*

# ROGER SHERMAN

*and the Independent Oil Men*

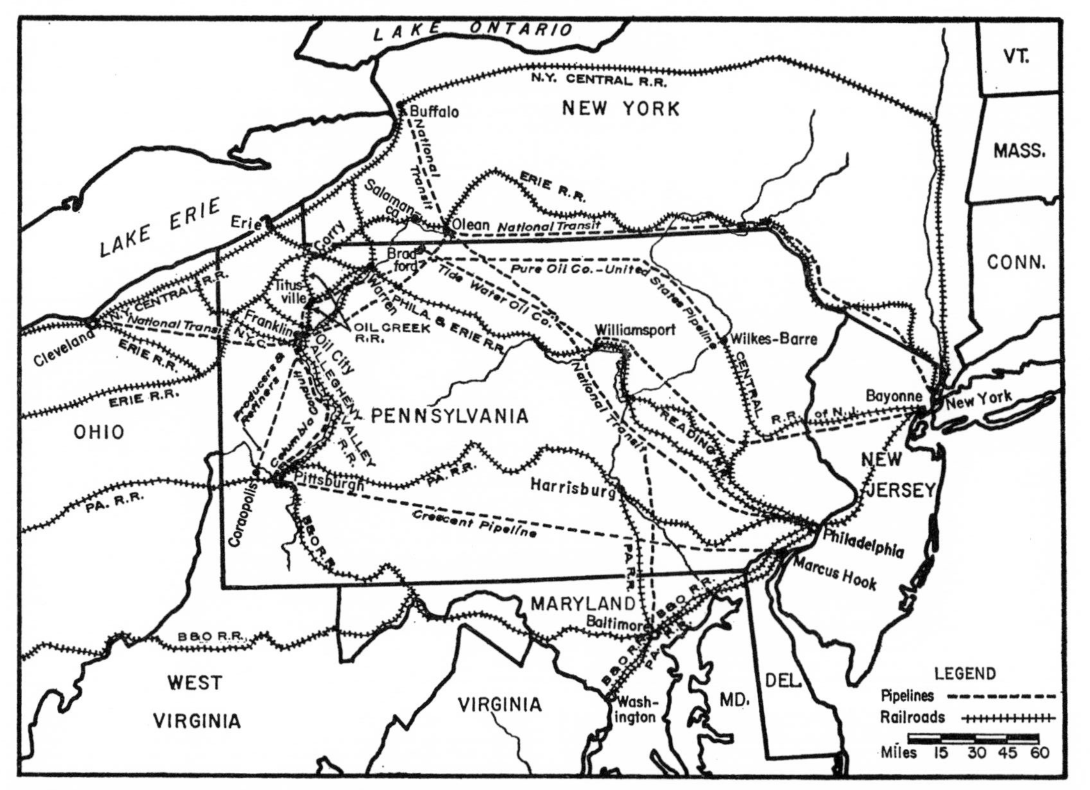


Pipelines and Railroads, 1896

# CHAPTER I

# *An Ex-Confederate Attorney in the Oil Region*

ROGER SHERMAN of Titusville, Pennsylvania, was born on July 28, 1839, at Randolph, Tennessee, a small town situated on the First Chickasaw Bluff upstream from Memphis on the Mississippi River. He was the only child of a doctor from upstate New York. His mother, Phoebe Conkling Sherman, a woman of refinement, calm dignity, and literary tastes, was the sister of a United States District Judge and an aunt of Roscoe Conkling, who became the influential United States Senator from New York.

Dr. Isaac DeBlois Sherman, Roger's father, a Williams College medical alumnus, was descended from the Sherman family of Connecticut. At Syracuse, New York, he had practiced medicine, become a merchant, and engaged in Jacksonian journalism. In 1833 he sold his newspaper and medical practice preparatory to removing to Texas. When an anticipated Mexican land grant there did not materialize, two years of travel in the West convinced him of the superior prospects of Randolph, and with his wife he moved there in 1835 to practice medicine. He fell in love with "the sweet southwest," as he named the region. At Randolph he soon purchased a large house on the bluff in the upper town. Seven years of profitable practice enabled him to purchase a stock-raising plantation of 1,200 acres across the river in Arkansas. In Randolph he aided the academy as treasurer and trustee. He helped the Episcopal church erect its edifice where his only child Roger was baptized by the Bishop of Tennessee and named after his distant relative, Connecticut's most distinguished signer of the Declaration of Independence.[1]

Had the river not swept away the lower town Roger might have grown up in Randolph, the son of a wealthy, public-spirited doctor, but the depopulation of the town obliged his parents to move in 1845. His father canvassed vainly the possibility of a medical practice in New Orleans before deciding to remove to their "Pecan Point" plantation on Mississippi Point, Arkansas, where he managed the plantation while practicing medicine in the area around Osceola.

Both parents lavished attention on their only son. His survival of a combined attack of whooping cough, malaria, and diphtheria in September 1844 was miraculous. He convalesced at Paducah, Kentucky, under his mother's care in the home of her cousin, the merchant Frederick Conkling.[2] A year later, as a large boy of six years, Roger was riding his pony about the Arkansas plantation and beginning his education by reading the New Testament and studying Webster's "Spelling Book."[3] When he was seven his father described him as "a fine noble-looking fellow, boisterous, & disposed to frolick and cutting, critical remarks."[4]

That year Mrs. Sherman took Roger to visit their New York relatives at Parma, near Rochester, and at Amagansett on eastern Long Island, her family's home. Because of her husband's desire that Roger receive formal schooling they remained at Amagansett a

[1] Roger Sherman Papers (Yale University Library), MS certified petition, I. DeBlois Sherman *et al.* to the President, Senate, and House of Representatives of the Republic of Mexico, May 3, 1833, I. DeBlois Sherman to Mrs. N. and Miss M. C. Sherman, July 12, 1836, October 24, 1837, May 10, 1838, printed announcement, Randolph Institute, January 27, 1840; Roger Sherman, *The Shermans: A Sketch of Family History and Genealogical Record, 1570–1890* (Titusville, 1890), pp. 22–27; Samuel P. Bates, *Our County and Its People: A Historical and Memorial Record of Crawford County Pennsylvania* (Meadville, 1899), p. 799.

[2] Sherman Papers, E. M. Conkling to I. DeBlois Sherman, July 9, August 22, September 1, 1844, I. DeBlois Sherman to E. M. Conkling, n.d., Mrs. I. DeBlois Sherman to Mrs. E. M. Conkling, August 6, 1844.

[3] *Ibid.*, "P. C. S." to E. M. Conkling, August 7, 1845.

[4] *Ibid.*, I. DeBlois Sherman to Mrs. M. E. Owen, January 2, 1847.

year, returning then to Parma for a year's visit with Mrs. Sherman's brother, Mr. Eleazer Conkling. Mrs. Sherman wanted her husband to sell "Pecan Point" and his medical practice and remove to a more promising location, but plantations in Arkansas were unsalable during the Mexican War.[5]

In 1849 Dr. Sherman purchased a house in Paducah, to which he took his family. Roger attended the academy there for three years where records show that in his eleventh year he ranked first or second in all his studies. After another year at the academy he went with his mother to Geneva, New York, where he attended the Rev. Dr. Prentice's preparatory school in preparation for entering Harvard College. After two years his father became unable to finance his education further and at fifteen years of age Roger was suddenly obliged to support himself.[6]

Roger Sherman's first position was that of chainman in a civil engineering party working on the enlargement of the Erie Canal. When a political revolution at Albany ousted the engineering staff, two of the discharged engineers suggested to him that they go to Iowa to survey the right of way of the Burlington and Missouri Railway. He became chainman and then leveler of that surveying crew. Incapacitated by carbuncles on a knee during the first winter he went to his mother's relatives at Parma to convalesce. While there he learned that a fire had destroyed his father's plantation home together with his library and deeds. Discouraged though he was, Dr. Sherman advised him to "live pure & holy" as his mother had taught him—she had died in 1855. "One bad habit blights a noble life," his father said, and added: "The obscene, the immoral, the scoffer insidiously, step by step, poisons many a youth & brings him to shame & disgrace."[7]

After recovery Roger Sherman rejoined his surveyors' party and

[5] *Ibid.*, "P. C. S." to I. DeBlois Sherman, July 23, 1846, May 17, 1847.

[6] *Ibid.*, I. DeBlois Sherman to Mrs. N. C. Owen, July 27, 1851; Sherman, *op. cit.*, pp. 28–29.

[7] Sherman Papers, T. U. Bates to Roger Sherman, June 28, 1856, G. W. Pierce to Roger Sherman, July 21, 29, December 28, 1856, B. F. Conkling to Roger Sherman, January 4, 1857, I. DeB. Sherman to R. Sherman, February 24, 1857; Sherman, *op. cit.*, pp. 29–31.

helped it to complete the preliminary survey of the Burlington and Missouri Railway. After the financial panic of 1857 obliged suspension of that construction he visited his father who had gone back to the Pecan Point Plantation, and then went to Parma where he began to read law, using his late uncle Nathaniel Conkling's Blackstone which his aunt Elizabeth Conkling gave him when he promised to join the bar. Dr. Sherman invited him to return to the plantation and complete his legal studies there.[8] He accepted, visited Randolph, and read law on the plantation, while corresponding with northern relatives at Parma and at Cambridge, Michigan. In a newspaper scrapbook he preserved clippings on public affairs.[9]

Pecan Point Plantation was south of Osceola on the bank of the Mississippi River. Although the house was on high ground, the lack of levees resulted in frequent flooding of the fields and pastures and of the land between it and Osceola. Dr. Sherman was frequently unable to attend meetings there, and Roger's bar examination in the court was delayed by the flooding and by a bout with malaria. Finally, however, on November 7, 1860, he was admitted to the bar in Osceola by Judge Earl C. Bronaugh. Although a flood prevented Roger Sherman from attending the circuit court, he immediately enjoyed a profitable practice among nearby planters.[10] Then, a few months after he began to practice, the Civil War "put an end to all business."

Both father and son openly opposed the secession of Arkansas. They desired the peaceable extinguishment of slavery and rejected the doctrine of secession. As maligned "Union Savers" they regarded the secessionists and the "Black" Republicans both as treasonable conspirators. However, during the excitement precipitated by the firing on Fort Sumpter Roger Sherman was persuaded that this resulted from northern aggression and concluded that the South must resist. After Arkansas seceded, the burgeoning war spirit

[8] Sherman, *op. cit.,* p. 32.

[9] Sherman Papers, MS Newspaper Scrapbook.

[10] *Ibid.,* MS Memorandum by I. DeB. Sherman, License from Judge Earl C. Bronaugh to Roger Sherman, First Judicial District, Arkansas, November 7, 1860; Sherman, *op. cit.,* p. 40.

helped to persuade the young lawyer that he should fight for his region.[11] In January 1862 he purchased a horse and informed his father of his decision. Dr. Sherman replied: "If you think it your duty, my son, go." Roger made his way to Memphis, where he enlisted for twelve months in Company D of Major Nathan B. Forrest's Tennessee Cavalry regiment.[12]

His reminiscences, published in *The Shermans* in 1890, contain a vivid account of his participation in the defense of Fort Donelson and of the escape and retreat of Forrest's cavalry to Nashville and then to Murfreesboro. There he wrote his father that he was "still well" and a guest at Captain Fletcher's quarters in General Hindman's command. "Universal indignation is felt in the army at the desertion of Nashville," he reported, and attributed the loss of Fort Donelson to its small, unreinforced garrison. Forrest's regiment had shrunk to 292 out of 750 with which it had arrived at Fort Donelson.[13]

After recruitment of new men for the regiment during two weeks at Huntsville, Alabama, Sherman marched with it to join General Albert Sidney Johnston's army at Burnsville, Mississippi. With his company he scouted in advance of its march toward the Tennessee River. They reached the vicinity of Shiloh on April 5, and participated in the great battle of the following two days, sharing in the victory of the first, on the second being driven in hard fighting by the reinforced Union Army back over the ground won the day before. As courier for General Hardee, Roger Sherman had an excellent opportunity to observe the progress of the battle. Returned to his regiment which was assigned to General Breckenridge's division covering the rear of the retreating Confederates, Sherman engaged in incessant skirmishing.

[11] Sherman, *op. cit.*, pp. 40–47; Bates, *op. cit.*, p. 800.

[12] Sherman, *op. cit.*, p. 47; Bates, *op. cit.*, p. 800; Confederate States of America, *Military Archives* (National Archives, Washington, D.C.), Muster Roll of Company D, Third Tennessee Cavalry.

[13] Sherman, *op. cit.*, pp. 47–53; Thomas Jordan and J. P. Pryor, *The Campaigns of Lieut.-Gen. N. B. Forrest, and of Forrest's Cavalry* (New Orleans, n.d.); Sherman Papers, R. Sherman to Dr. Sherman, February 24, 1862.

Picket and scouting duty became severe. On October 6, near Pocahantas, Tennessee, he was captured and imprisoned briefly in a vermin-infested courthouse. Released on parole he returned to his regiment when exchanged to find that food was more scanty and clothing more difficult to obtain. Confederate money was of diminishing value. "The southern soldier had little to eat or wear, and in the southwest the Confederate armies were continually beaten." It was a discouraging experience, as Arlin Turner, biographer of George W. Cable, demonstrates. Sherman lost two horses in action, had one stolen, a fourth was "used up," and he was obliged to buy replacements on credit. These debts he faithfully repaid during Reconstruction.[14] Despite numerous responsible assignments and personal association with officers he remained a private during his military career.

His father remained behind on his plantation, his colored servants leaving him to go to Fort Pillow some six miles distant where they reported early in February 1863 that Dr. Sherman possessed a large hoard of gold. This was untrue but was believed, and shortly afterward three men in Federal uniforms were seen in the area. Three weeks later neighbors found the house open, looted, abandoned. A quarter of a mile away they found Dr. Sherman bound to a tree, only recently dead, having survived torture and injury to perish alone shortly before he was found. All moveable property had been stolen from the house leaving the plantation "a little less than worthless." Dr. Sherman's fate and the looting of his plantation illustrate the violent plundering that was experienced by the conquered portions of northeast Arkansas and west Tennessee behind the Union lines.[15]

Some months later, probably after the Vicksburg campaign, Roger Sherman became separated from Forrest's corps and made his way northward to his relatives at Cambridge, Michigan.[16] The claim in his reminiscences that he had fought in all the battles of

[14] Sherman, *op. cit.*, pp. 53–60; Arlin Turner, *George W. Cable* (Durham, 1956), pp. 21–34; Bates, *op. cit.*, p. 800; Sherman Papers, R. Sherman to Rev. Kendrick Metcalf, April 10, 1871; Confederate States, *Military Archives*, Muster Roll, Company 9, Third Tennessee Cavalry.

[15] Sherman, *op. cit.*, pp. 59–60.

[16] *Ibid.*, p. 63.

the war in the southwest is unacceptable. The careful account of his career in Titusville, Pennsylvania, published in Samuel P. Bates' *Our County and Its People* states frankly that he arrived in the north in 1863.[17]

Sherman went next to Parma, New York, to visit with his uncle Eleazer Conkling and to resume the study of law. From there he removed to Erie, Pennsylvania, before January 1865, where he worked as a lawyer's assistant pending admission to the bar. He was also editor of the weekly *Erie Observer,* a Democratic organ published by another lawyer, Benjamin Whitman, who became his lifelong friend. Sherman became sufficiently well known to be able to secure from Congressman G. A. Scofield a letter of introduction that enabled him to visit his plantation in Arkansas.[18]

His Conkling relatives were doubtful of the wisdom of the trip. Friends of the Erie bar thought otherwise and gave him letters of introduction to officers of the military government. Allan A. Craig introduced him to Major J. A. Nunes of Louisville as "a member of our bar, and a gentleman. Use him as you would me, under the same circumstances." [19] Sherman had become interested in oil production as a field of speculative investment during a visit to the Oil Region of northwest Pennsylvania. Already some were thinking that oil might be found elsewhere. Incidental to his trip to Pecan Point Plantation was a search for oil lands in eastern Kentucky and Tennessee. As he traveled he sent letters reporting his observations to the *Erie Observer.*[20]

Upon his return to Pecan Point Plantation he gathered together the few remaining books and dug up some silver spoons. Moreover he ascertained that he would be unable to farm the 1,200 acres for some years because of the demoralized labor supply. A law practice

[17] Bates, *op. cit.,* p. 800.

[18] *Ibid.,* p. 800; Sherman Papers, E. M. Conkling to R. Sherman, December 11, 1864, G. A. Scofield to ?, January 30, 1865.

[19] Sherman Papers, A. Conkling to E. M. Conkling, February 16, 1865, A. A. Craig to Major J. A. Nunes, May 3, 1865, James Sill to Major A. A. Craig, February 8, 1865, J. A. Galbraith to R. Sherman, December 11, 1864.

[20] *Ibid.,* MS "Memorandum of Kentucky Oil Lands," E. M. Conkling to R. Sherman, June 12, 18, 1865, R. Sherman to A. Matson, June 15, 1865.

at Osceola was equally out of the question. He decided that with his "tastes" he would be unhappy there. His rescue of an old lady and her sick daughter from a squad of colored Union soldiers that were looting their house in Tipton, Tennessee, convinced him that property in the area was unsafe, and he returned to Erie in mid-June 1865.[21]

In July, backed by James Sill and other attorneys, he applied to the president judge of the circuit court, Samuel P. Johnson, for admission to the bar. Certain other lawyers maintained that he was disqualified because he had been a "rebel." The judge admitted this objection as decisive and rejected Sherman's petition.[22]

He left for Pithole, Pennsylvania. That turbulent boom town was situated southeast of Titusville on a hill in Venango County near the famous United States Well. Already the immigration of producers, oil-well workers, merchants, hotelkeepers, artisans, speculators, and the curious had brought 5,000 residents to drill for oil and to do business there. Hastily built structures lined streets in which occasional stumps and rocks protruded from the almost fluid mud during the rains. On July 19 Sherman opened there a branch law and real-estate office for Clark Ewing of Titusville.[23]

Already Sherman had caught the "oil fever," although, because of temporary waning of the speculative boom in Oildom, he had been unable to exploit some oil leases he had secured while exploring mountainous Kentucky and Tennessee.[24] G. W. Waite, a friend from Cleveland, Ohio, learned that he was dabbling in "ile." "Well my good fellow," he wrote Sherman, "I sincerely hope you will make your heap, and not become one of the 'unfortunates.' "[25]

[21] *Ibid.,* E. M. Conkling to R. Sherman, June 12, 1865; Sherman, *op. cit.,* pp. 61–62.

[22] Sherman, *op. cit.,* pp. 62–63.

[23] *Ibid.,* pp. 63–64; "Crocus" (Charles C. Leonard), *The History of Pithole* (Pithole, 1867), *passim;* Paul Giddens, *The Beginnings of the Oil Industry* (Harrisburg, 1941), p. 116, quoting *Oil City Register,* June 1, 1865.

[24] Sherman Papers, MS "Memorandum of Route," MS "Memorandum of Kentucky Oil Lands," R. Sherman to A. Matson, June 15, 1865.

[25] *Ibid.,* Waite to Sherman, June 25, 1865.

The rush to Pithole was still in progress when Sherman arrived. Capital flowed in for investment in oil lands, leases, producing properties, service enterprises, and the swift construction of a clapboard city housing the 15,000 who settled in the town by September. Many thousands more came, saw, and speculated in oil or in the stock of ephemeral oil-producing companies, and left, some sadder and wiser, others enriched by the golden flood. Before autumn the two banks, fifty hotels, the *Pithole Daily Record,* the plank road to Titusville, a water works, a post office with a business volume ranking third in Pennsylvania, a school system, and two churches presented an appearance of permanence. Waves of excitement attendant upon each new oil strike and the frenzied speculation in town lots, crude oil, shares of oil companies, and the surrounding oil lands or leases of them, set the tone of the new city. The dearth of entire families and of women led to greater respect for prostitutes than was customary east of the Rocky Mountains, while vice, crimes of violence, gambling, and alcoholism flourished in a manner that provoked widespread comment.

The *Nation* and the eastern press gave due publicity to the gushing United States Well and the Twin Wells, the profane teamsters with their hundreds of wagons hauling in equipment and hauling barrels of oil away to Titusville, the forest of derricks, the railroad being laid to Oleopolis nearby on the Allegheny River, the stream of immigrants walking and riding into the city, the unpainted United States Hotel, the $80,000 Bonta House, and the Danforth House erected on land leased for $14,000 a year. The bad meals, bad liquor, bad table manners, sallow complexions, and unusual intelligence of the men also provoked comment.[26]

Astute observers noted that most of the capital that financed Pithole's spectacular boom was coming from the eastern cities. The

[26] *Nation,* I (New York, September 21, 1865), 370–372, quoted in Paul H. Giddens, *Pennsylvania Petroleum, 1750–1872* (Harrisburg, 1947), pp. 284–292; "Crocus," *op. cit., passim;* "Petroleum, Chapter VI," *Petroleum Age,* I (Bradford, May 1882), 186–188, (June 1882), 219–222; Paul H. Giddens, *Early Days of Oil* (Princeton, 1948), pp. 60–65; Harold F. Williamson and Arnold R. Daum, *The American Petroleum Industry, 1859–1899* (Evanston, 1959), pp. 122–124.

oil mania reached an unparalleled peak during the years from 1865 to 1867. By the later date the capital invested in the petroleum industry alone in the Oil Region was estimated to be $331,600,000. Of this sum Philadelphia invested $168,715,000 and New York City $134,000,000. Titusville enterprisers supplied $5,000,000 and Pittsburgh $15,740,000. Cleveland's $2,200,000 seemed puny indeed.[27] This situation produced outside control not only of oil-producing properties but also of railroads, banks, telegraph and insurance companies, and a score or more of refineries that processed crude petroleum on Oil Creek. It was inevitable that Philadelphia and New York capitalists should rival one another in attempts to dominate the Oil Region.

The few fragments that survive of his papers of that period show that Sherman was obliged to pay $2,000 a year rent for a two-room upstairs office suite. He drafted bills of sale for fractions of the oil produced upon a piece of real estate. He drafted oil leases on more distant properties as prospecting expanded.[28] The price of oil was fluctuating wildly, rising abruptly with the completion of the plank road to Titusville and the railroad to Oleopolis and then sinking with overproduction to $1.35 a barrel in December 1866. Not until 1869 would crude petroleum rise to $7 a barrel to give renewed impetus to the great oil boom.[29] While men sat on logs under the hot sun, trading oil properties and leases, Sherman sweated in his office as Ewing's employee and longed for admission to the Venango County bar so that he could practice law on his own initiative. His reminiscences pass over his Pithole experience briefly.

As a war veteran he probably was not shocked by the shooting affrays and fisticuffs, the loud revels, the extraordinary gambling, the "social crime" on a scale matching that of ancient Athens. Coming from a family possessing ethical standards he was undoubtedly stunned at the open fleecing of investors, the open operations

[27] "Capital in the Oil Regions," *Petroleum Age,* I (December 1881), 35.

[28] Sherman Papers.

[29] George Ward Stocking, *The Oil Industry and the Competitive System. A Study in Waste* (Boston, 1925), pp. 10–11.

of highwaymen and gangs of outlaws. His Episcopalian background drew him into the ranks of the better citizens who organized a vigilante committee and secured a borough government that closed the "dens of infamy" and brought security to property in September 1865.[30] Repelled by the "oil smellers," geology "professors," and liars who imposed upon newcomers, Sherman witnessed the shameless profiteering that disclosed how tarnished business ethics had become in the rush for maximum profits.

Terminating the association with Ewing, Sherman established his own law office in December 1865. He took into partnership a former Union Army lieutenant, Theodore C. Spencer, who attended in Franklin to the courtroom aspects of their practice. Then Sherman's second application for admission to the Erie bar was rejected by Judge Johnson because of his previous Confederate military record. This was criticized severely by A. D. Wood of Warren, who became Sherman's firm friend. Encouraged by Wood, he hoped that "common sense if not common justice" would remove the ban barring him from the courts despite the deep anti-Confederate prejudice of the city's Unionists.[31] His generous uncle Eleazer Conkling hoped that he would strike "il" and recover from the "blues." He relished Sherman's descriptions of Pithole.[32]

Although it was humiliating to be obliged to rely upon Spencer to handle their firm's courtroom business, Sherman learned much of oil and real-estate law and of Pennsylvania legal procedures during 1865 and 1866. He began to purchase a law library. He became quickly an adviser on investment opportunities. M. M. Wallace of Erie complimented him on his analysis of the limits of the Oil Region and of the probable "direction of the excitement." Sherman described to other Erie attorneys how all the land in Pithole was held subject to leasehold, and said that neither mechanics erecting buildings and derricks nor merchants could secure themselves by

[30] "Petroleum, Chapter VI," 186–188, 219–222.

[31] Sherman Papers, A. D. Wood to Sherman, October 12, 1865, C. E. Ewing to Sherman, September 13, 1865, F. B. Guthrie to Sherman, December 6, 1865; Sherman, *op. cit.*, p. 64.

[32] Sherman Papers, E. M. Conkling to Sherman, July 20, 30, August 20, 30, September 7, October 15, 1865.

taking leasehold mortgages. They agreed that a mechanic's lien was not available but advised him that common law and a Pennsylvania statute of 1865 authorized mortgaging of mining leases. James Sill of Erie, while congratulating Sherman on his "excellent position & prospects at Pithole," agreed that a new local law applicable to the Oil Region counties might be secured. Ultimately Sherman drafted such a bill and secured its enactment in 1868. This authorized a lien upon oil-mining leaseholds and protected the laborers from the numerous frauds practiced hitherto upon them.[33] This was Sherman's first attempt to cope effectively with the unethical practices of Oildom.

Sherman's association with Spencer drew the sting from Radical Republican criticisms. His friends of the Erie bar sent them business, offered assistance, and encouraged Sherman to keep alive his application for admission to the bar. As the boom tapered off, Sherman accepted the advice of an Erie friend that he apply for admission elsewhere. He turned to a Democrat, President Judge John Trunkey of Mercer and Venango Counties, and was informed that he would be admitted to practice in Venango County immediately. After a formal examination by a committee of its bar he was admitted to practice on February 22, 1867.[34] Sherman and Judge Trunkey remained friends for many years. Following this admission to the bar Sherman's and Spencer's practice improved. However, Pithole's recovery from a succession of disastrous fires was but partial.

Sherman and a young client, Elisha G. Patterson, became warm friends, and for the next fifteen years Sherman was his attorney. In 1867 Sherman became President of the Pithole City School Board. He purchased the building housing his law office. Apparently he was able to engage in oil production—since he was listed second at a reunion of the "Pithole Pioneers" in 1890—although he had been unable to sell Pecan Point Plantation to raise investment capital.

[33] *Ibid.,* M. M. Wallace to Sherman, October 2, 1865, Sherman to Sill, October 3, 1865, Sill to Sherman, October 6, 1865, Sherman, *op. cit.,* p. 71.

[34] Sherman Papers, J. A. Galbraith to Sherman, December 11, 1865, B. Whitman to Sherman, December 10, 1865, MS license to practice law in Venango County, March 1, 1867; Sherman, *op. cit.,* pp. 64–65.

Another of the Pithole group with whom he became associated professionally during the years from 1865 to 1868 was Lewis Emery, Jr., who became a lifelong friend. For long periods Sherman was his counsel and business adviser.[35] Much of the history of the long conflict between the independent oil producers and business monopoly in the Oil Region before 1880 would derive from the friendship of Sherman with Patterson and Emery.

Sherman was ruined financially in late February 1868 when a great fire destroyed most of Pithole, and with it his law office, papers, the building housing them, and all but a few of his books. When he presented his fire-insurance claim to the Western Phoenix Insurance Company of Chicago he was glibly informed that the company had been liquidated a few weeks earlier. So many of Pithole's poverty-stricken men and women went to Titusville that Horace Greeley went there to lecture for their benefit. Sherman found himself virtually penniless, but he was determined not to succumb to misfortune. He agreed undoubtedly with his uncle Eleazer Conkling's quotation from Robert Burns: "Fortune is but a bitch." He remained in Pithole and continued his law practice for some weeks. He secured legislation desired by the school board. Despite attempts to rebuild the town, Pithole was doomed by the rapid decline in production of its wells owing to flooding with surface water.[36]

Sherman and Spencer left Pithole in late March for the new producing district at nearby Pleasantville arriving there on April 1, 1868, in the midst of a rush that transformed the village into a town of 3,000. They formed a partnership with M. C. Beebe, a

[35] Sherman Papers, Record of E. G. Patterson *v.* Tide Water Oil Company, E. G. Patterson's testimony, December 20, 1882 (copy); Elisha G. Patterson Newspaper Scrapbook (courtesy of Mrs. Eleanor F. Schreck, Independence, Kansas), for the 1890 clipping of the reunion of the "Pithole Pioneers"; *First Annual District Report,* Pithole City School Board, July 3, 1867.

[36] Sherman Papers, E. M. Conkling to Sherman, March 1, 1868, P. L. Sherman to R. Sherman, March 7, 1868, A. P. Duncan to R. Sherman, March 5, 9, April 4, 1868; *Titusville Morning Herald,* March 3, 1868; Sherman, *op. cit.,* p. 71.

local lawyer. Within a little more than a year their practice extended to the entire field of active oil production in Venango, Crawford, Warren, and Forrest counties. At the same time Pleasantville was acquiring two banks and an opera house to provide the service and recreational amenities that Oildom required.[37]

Three weeks after his arrival at Pleasantville Sherman became the counsel of Samuel Van Syckel, the inventor, builder, and operator of the first successful petroleum pipeline in the Oil Region. Sherman had known him at Pithole when Van Syckel pumped oil through the line from the wells there to Miller Farm on the Oil Creek Railroad. Built with $30,000 borrowed from the First National Bank of Titusville, Van Syckel's wrought-iron welded pipeline had worked successfully as early as October 1865 and had been very profitable, grossing $2,000 a day from its charge of $1 a barrel. Embarrassed by the failure of an important debtor the First National Bank demanded immediate repayment of the loan to Van Syckel. When he defaulted, it foreclosed and then sold the pipeline to a buyer who promptly marketed it to William H. Abbott, rising Titusville capitalist and oil producer who was a stockholder in the bank.

In 1867 Abbott merged the Miller Farm pipeline with another pipeline which was operated between Bennehoff Run and Shaffer Farm on the Oil Creek Railroad by Henry Harley, an engineer graduate of Renssaeler Polytechnic Institute. They incorporated the resulting company as the Allegheny Transportation Company. They absorbed other competing pipelines and extended their system as the producing area expanded. By January 29, 1868, when Harley was elected President, the Allegheny owned or controlled nearly all the pipelines on Oil Creek. This dominant position, which had displaced the teamsters' transportation of oil from the wells to the railroads and refineries, had been financed partially by Jay Gould,

[37] J. T. Henry, *The Early and Later History of Petroleum* (Philadelphia, 1873), pp. 590–591; Giddens, *Early Days,* pp. 85–87; Williamson and Daum, *op. cit.,* p. 130; Sherman Papers, enclosure, Sherman to John Livingston, July 1, 1869.

President of the Erie Railroad, and James T. Fisk of New York, his close associate, who desired to capture the oil traffic for the Erie Railroad.[38]

Van Syckel wanted Sherman to recover his pipeline for him. He also sought help from Henry C. Ohlen, a New York City oil buyer who had been his partner in building the Miller Farm pipeline. Although Van Syckel demonstrated that five of the six directors of the bank were stockholders in his company, the litigation was unsuccessful.[39] For Sherman the chief result was to bring him into close quarters with long-faced, astute William H. Abbott.

The Allegheny Transportation Company played the local railroads off against each other. In return for delivering oil to them it exacted secret payments inaccurately termed rebates or pipage. Among the railroads, the Oil Creek Railroad, controlled by Philadelphia capitalists associated with the Pennsylvania Railroad, achieved a monopoly position by absorbing the Warren and Franklin and the Farmers' railroads on Washington's Birthday, 1868.[40] Reincorporated as the Oil Creek and Allegheny River Railroad, Abbott, Harley, and Gould took especial measures to ensure that oil would move over it in greater quantity to the Erie Railroad rather than southward to Pittsburgh and Philadelphia, cities whose natural advantages as refining and export centers appeared to be obvious.

Gould erected a large refinery with oil docks at Weehawken, New Jersey, which he leased to the Erie Railroad. The oil-buying and shipping firm of Bostwick & Tilford of New York City was then allowed to sublease the Weehawken terminal and docks and

38 *Titusville Morning Herald,* February 1, 1868; *Titusville Daily Courier,* October 11, 1870; Andrew Cone and Walter R. Johns, *Petrolia: A Brief History of the Pennsylvania Petroleum Region* (New York, 1870), pp. 113–114; N. M. Allen, "Petroleum and Our Connection Wherewith," in Bates, *op. cit.,* pp. 415–417; "Crocus," *op. cit.,* pp. 37–38; Henry, *op. cit.,* pp. 283, 368–369, 527–532; Arthur Menzies Johnson, *The Development of American Petroleum Pipelines* (Ithaca, 1956), pp. 6–14.

39 Sherman Papers, S. Van Syckel to Sherman, April 21, 1868; *Titusville Morning Herald,* December 11, 1868.

40 *Titusville Morning Herald,* February 25, 1868.

was granted exclusive large drawbacks on oil freight rates from Titusville to that terminal. Gould made Henry Harley the "oil agent" of the Erie Railroad with control over freight rates on crude-oil and refined-oil products, and of the leased Atlantic and Great Western Railroad. Its receiver, like Gould, owned stock in the Allegheny Transportation Company.[41]

Harley then invited three small refining companies of Cleveland, Ohio—Rockefeller, Andrews & Flagler, Clark Payne & Company, and Westlake Hutchins & Company—to accept together gratis a fourth interest in the Allegheny Transportation Company and to receive large secret drawbacks on oil rates from Titusville to Cleveland on the Erie-Atlantic and Great Western system. This arrangement was made by six contracts drafted by as many lawyers. Roger Sherman drafted the railroad rate drawback contract between the Atlantic and Great Western Railroad and the three Cleveland refining companies. He retained this after its signing, together with the other five contracts and a letter from Gould to Robert B. Potter, the Atlantic and Great Western's receiver.[42]

The secret rate drawbacks of this triple alliance of the pipeline monopoly, the Erie-Atlantic and Great Western system, and the three Cleveland refiners derived from the general practice of paying commissions to agents to secure traffic for the railroads. It was Gould's and Potter's purpose, however, to build up the traffic to Cleveland and New York City by large exclusive drawbacks, which was an indefensible innovation by common carriers. In each instance the beneficiaries received a decisive advantage in competition that stimulated a precipitate expansion in their business volume. It accelerated Cleveland's rise as a refining center so swiftly that it won primacy from Pittsburgh by 1871. The intervention of Jay Gould's Erie Ring of New York City into the petroleum industry affected decisively the rivalry of those two midwestern cities while

[41] Julius Grodinsky, *Jay Gould: His Business Career 1867–1892* (Philadelphia, 1957), pp. 59, 69, 82–83.

[42] Chester McArthur Destler, "The Standard Oil, Child of the Erie Ring, 1868–1872, Six Contracts and a Letter," *Mississippi Valley Historical Review,* III (June 1946, March 1947), 89–114, 621–628.

accelerating the rise of New York City as a refining and exporting center at the expense of Philadelphia.

Meanwhile, upon learning in August 1868 that Sherman had his "hands full," his friend, E. L. Keenan, a Philadelphia attorney, congratulated him and hoped he would "realize sufficient to enable you to start well elsewhere."[43] After Spencer withdrew from the firm, Sherman and Beebe moved their office to the Chase House in Pleasantville where they rented elegant quarters for $500 per annum, and also opened a branch office in Franklin. The firm announced that it gave special attention to "business interests of parties residing abroad, and to the examination and perfection of Land Titles." Sherman was again attending to the interests of investors in oil properties. On December 26 he repaid the balance due on a loan that an Erie friend had extended to him in February 1865.[44]

He was now consulted by producers on how to invoke his new mechanics lien law to protect themselves from swindling contractors. He protected a Titusville investor by examining the records of the McClintock Farm & Cherry Tree Petroleum Company of Franklin. He consulted A. P. Duncan on the merits of a bill to prevent the flooding of oil wells. He began again to purchase a law library. Working hard he reported his professional success to his uncle Eleazer Conkling who termed it "most extraordinary," and sent him an "Ambrotype" of his mother, terming it "a very perfect likeness."[45] This delighted the ex-Confederate, who was becoming interested in Miss Alma Seymour, daughter of Claudius B. Seymour, an old resident of Pleasantville who had become involved in oil production.

Sherman's professional success was evinced further when The Mercantile Agency, E. C. Dun & Company, made his firm its agent for Venango County as the source of confidential data on the credit rating of commercial establishments. J. M. Bradstreet & Son of New

[43] Sherman Papers, Keenan to Sherman, August 27, 1868.

[44] *Ibid.,* MS statement, enclosure, B. Whitman to Sherman, December 29, 1868, W. A. Galbraith to Sherman, December 24, 1868.

[45] *Ibid.,* E. M. Conkling to Sherman, November 15, 1868, A. P. Duncan to Sherman, February 12, 1869.

York City did so likewise. Both appointments aligned Sherman permanently with those who were working to enforce adequate standards of business ethics in the Oil Region. Then, on October 18, 1869, he was admitted to practice before the Supreme Court of Pennsylvania, Western District.[46]

The decline of the Pleasantville oil field and desire for a larger practice led Sherman and Beebe to move their main office to Titusville on July 5, 1870, immediately after Sherman had been admitted to practice in Crawford County. In October he was admitted to practice in the United States District and Circuit Courts for Western Pennsylvania.[47] This placed him in the front rank of the bar of the Oil Region. Relocation in Titusville, the most stable business center of the Oil Region, now presented him with an opportunity to develop a permanent practice and accumulate a competence. Sherman and Beebe erected a three-story wooden business building to house their office and provide rental property. There Roger Sherman maintained his office for the remainder of his life.[48]

Sherman directed the firm's Titusville office. Most of its business lay outside the realm of criminal law, although Sherman did appear in such cases in justice and Superior courts. Civil actions, searching titles, drafting deeds, conveyances, mortgages, and oil leases, and drawing up contracts in the fields of oil producing, marketing, and transportation were the largest part of its work. Sherman became a recognized expert in equity jurisprudence, frequently arguing cases on appeal in higher Pennsylvania courts and occasionally before the United States Circuit Court.[49] Litigation of this type acquainted him with the economic life of the Oil Region. A survey of forty per cent

[46] *Ibid.,* E. C. Dun & Company to Sherman & Beebe, n.d., J. M. Bradstreet & Son, "Instructions to Correspondents," MS certificate Thomas J. Keenan, Prothonotary, Supreme Court of Pennsylvania, Western District, October 18, 1869.

[47] *Ibid.,* MS certificate, John F. Morris, Prothonotary, Crawford County, April 12, 1875, MS certificate, S. C. McCandless, Clerk, United States District Court, Western Pennsylvania, October 18, 1870.

[48] *Ibid.,* Sherman & Beebe to Williams & Murray, February 4, 1872, Sherman to Beebe, May 31, 1875.

[49] *Ibid.,* Sherman to W. W. Darne, August 6, 1872.

of the appointment docket of the Crawford County Court of Common Pleas for January 1873 disclosed that the firm of Sherman & Beebe were scheduled to appear in sixteen actions, many of them in equity.[50]

As was customary for leaders of the bar, Sherman accepted a law student, M. J. Heywang, who read law with him from 1872 to 1875, was clerk of Sherman & Beebe, and remained his friend and admirer after becoming a practitioner of high professional standards.[51] Continual additions to Sherman's professional library facilitated such instruction. He added classics in other fields to his personal library in which he read for an hour daily before dinner. He was reading systematically outside the law in an attempt to acquire the cultural education of which he had been deprived when his father's financial embarrassment prevented him from attending Harvard College. His interest in Sharswood's *Ethics* derived from problems presented by the speculative society of Oildom. This explains his cutting reference to the defective federal bankruptcy act as the "Act entitled an act to hinder delay and defraud creditors, and encourage dishonesty conspiracy and fraud."[52] In 1874 he made the closing argument on the invitation of the prosecution in the sensational Hunter Well conspiracy case before "Justice Tracy's court" in Titusville. "His effort was a very able one and was listened to with close attention," the *Titusville Daily Courier* reported.[53]

Sherman's attitude toward the low moral standards of oil towns was revealed when a tenant of his and Beebe's office building allowed "persons not tenants of these rooms to visit them with prostitutes of the lowest kind and disturb other occupants of the building and disgust the neighborhood." He informed the tenants bluntly that he and Beebe were determined to "purify the moral

[50] Crawford County, Pennsylvania, Court of Common Pleas, *Appointment Docket,* LI (Courthouse, Meadville; January 1873).

[51] Sherman Papers, Heywang to Sherman, May 19, August 23, 1873; Bates, *op. cit.,* pp. 428–429.

[52] Sherman Papers, Sherman to McDevitt Campbell & Company, August 13 [1873].

[53] June 30, July 1, 1874.

atmosphere of our premises" and threatened "to take effective means to rid ourselves of this nuisance, if it occurs again." [54]

Sherman continued to serve as correspondent of Bradstreet & Son and of Dun & Company. They now required that he supply detailed data on the assets and credit standing of all types of business enterprises in Crawford County. Bradstreet & Son recommended Sherman & Beebe "at all times for the collection of claims." [55]

Sherman's income from his practice rose rapidly from $2,500 in 1870 to an average of $13,000 at the close of the decade. From 1870 to 1872 he derived additional income from his association with William H. Abbott, the Roberts brothers, John Fertig, and others in the founding of the *Courier* as the Democratic organ. As attorney for the paper's backers Sherman drafted the organization papers of the Titusville Printing Press Association, the publishing firm, and subscribed to sixteen shares of its stock. Abbott was President of the Association, Sherman the Secretary. In this manner Sherman became formally associated with the Abbott group of Titusville capitalists who in turn were allied with the Erie Ring. Colonel James T. Henry, editor of the *Courier,* publicized the activities of the Abbott associates, devoted much space to the Erie Railroad's next annual stockholders' meeting, and welcomed Jay Gould's attempt to rehabilitate that carrier. Henry also published accounts of the overthrow of the Tweed Ring, of the murder of Fisk, and of Gould's flagrant watering of the stock of the Erie Railroad in early January 1872. At the time this was published Abbott was nominated and then beaten in the Titusville Oil Exchange for offices from President down to the Committee on Room! [56]

[54] Sherman Papers, Sherman & Beebe to Williams & Murray, February 4, 1872.

[55] *Ibid.,* Sherman to Beebe, May 31, August 13 [1875], J. M. Bradstreet & Son to Sherman & Beebe, February 24, 1872.

[56] *Ibid.,* minutes, correspondence, bylaws, and contracts of the Titusville Printing Press Association; *Titusville Daily Courier,* October 1, 10, 14, 20, 1870, January 9, 1872; *Petroleum Age,* I (February, March, 1882), 97–98, 127; Bates, *op. cit.,* pp. 317–318; Henry, *op. cit.,* pp. 3, 360–374; Giddens, *Pennsylvania Petroleum,* pp. 254–256.

As an experienced journalist Sherman contributed political articles and other items to the *Courier* for three years. One of his efforts was a brilliantly successful campaign to elect Walter H. Lowrie, ex-Chief Justice of the Pennsylvania Supreme Court, as President Judge of the Crawford County Court of Common Pleas.[57]

Sherman did exhibit obtuseness toward the implications of the drawbacks that the Erie Ring awarded the Rockefeller group and Bostwick & Tilford. He also drew up contracts whereby the Oil Creek and Allegheny River Railroad paid drawbacks to the Allegheny Transportation Company—soon to be renamed the Pennsylvania Transportation Company. On March 19, 1870, he drafted the contract whereby this firm paid five cents a barrel for a year to the Standard Oil Company "on all crude moved for their use in the business of refining in Cleveland over the Atlantic and Great Western Railway." [58] Sherman may have recognized that this practice was an abuse. If so his association with the Abbott associates obliged him to smother his opinion.

The disillusionment of the Abbott associates with Gould and Fisk dated from as early as June 1871. The Abbott group sponsored construction of the Union and Titusville Railroad to defeat the Oil Creek and Allegheny River Railroad's monopoly. The construction contractors sold their half of the bonds and the common stock to Gould and Fisk in February 1871. Abbott did everything in his power to induce them to retain this. However, after they lost control of the Atlantic and Great Western Railway they sold their holdings in the Union and Titusville Railroad to the Oil Creek and Allegheny River Railroad. Then, in a renewed attempt to revive competition for Titusville's freight the Abbott associates purchased the construction contractors' holdings in the Titusville and Petroleum Center Railroad. It was obvious that they were not tools of the Erie Ring in matters affecting Titusville interests.[59]

[57] *Titusville Daily Courier,* October 1—November 9, 1870.

[58] Sherman Papers, MS contract between Allegheny Transportation Company and the Oil Creek and Allegheny River Railroad, October 8, 1868, MS contract between Allegheny Transportation Company and the Standard Oil Company, March 19, 1870.

[59] *Titusville Daily Courier,* January 22, 1872; Bates, *op. cit.,* p. 827.

Four days before Abbott explained his relation to these two local railroads Roger Sherman headed a group of seventeen citizens, including the Roberts brothers, H. C. Bloss of the *Titusville Morning Herald,* and N. M. Allen, a director of the *Courier,* who convoked a public meeting in Parshal's Opera House to further the completion of the Warren & Venango Railroad in which Cornelius Vanderbilt was interested. The *Courier* editorial on the meeting declared that the city was suffering from "excessive freight rates" and from railroad discrimination against the local refiners and approved of building another railroad to be part of another trunkline. Cleveland refiners, it declared, could ship crude oil to Cleveland, refine it, and then ship the refined product to New York City "at better margins of profit than can be done on the Creek." They paid nominal rates to New York City of 60¢ per barrel less than did Oil Creek refiners who, on completion of the Warren & Venango Railroad, would be able to operate at a profit "and trade of every description here will take on new life." Only the Abbott associates, from whom Sherman had withdrawn, protested.[60]

Titusville had good reason to be concerned at Cleveland's swift rise in oil refining. After it had been favored from 1868 to 1870 by the Erie-Atlantic and Great Western system, the Lake Shore & Michigan Southern Railroad, leased by the New York Central Railroad, constructed a branch line from Cleveland to Franklin in the Oil Region. Before its completion the Lake Shore granted the Standard Oil Company of Ohio a large secret drawback on the crude-oil rate to win this trade from the Atlantic and Great Western. The competing railroads paid the Standard Oil the contracted drawbacks during the rate war that ensued. In this manner Rockefeller played the Lake Shore against the Atlantic and Great Western to his own great advantage, while Cleveland enhanced the new primacy in oil refining that it had won with the aid of the Erie Ring. The Pennsylvania Railroad attempted to meet the new situation with equivalent oil-rate reductions and by large secret drawbacks to leading refiners in Pittsburgh and Philadelphia. The new advan-

[60] *Titusville Daily Courier,* January 18, 22, 1872.

tages to the four large competing cities were such that the oil refiners of the Oil Region were at a serious disadvantage in market competition despite their location close to the wells and closer to the seaboard than Cleveland and Pittsburgh.[61]

Sherman's leadership in the attempt to persuade Vanderbilt to complete the Warren & Venango Railroad to Titusville was an attack upon the Erie and the Oil Creek and Allegheny River railroads. It was also an attack upon the position of the Abbott associates who were still loyal to the Erie Railroad despite Gould's indifference to Titusville. Thus Sherman declared publicly his independence of the Abbott associates and their alliance with the Erie Ring. He emerged as a civic leader of influence who was determined to foster the city's welfare by an alliance with Vanderbilt. A year later, however, the Citizens Corps in which Abbott and Harley figured invited him to their first Inaugural Ball.[62]

Sherman's leisure during early 1871 had been devoted to his courtship of Alma Seymour. On March 16 he returned to Pleasantville to marry her. The announcement of this event provoked congratulations from local friends, professional associates, and his relatives. Pleased by a congratulatory letter from the Rev. Kendrick Metcalf of Geneva, New York, a friend of his mother when he studied in prep school there, Sherman replied: "After much poverty and a hard struggle, I have succeeded in securing a lucrative practice in my profession which leads me to hope that the wasted & unfortunate years of my early manhood, may in a measure be recompensed by success in my future life." Happily married he enjoyed his Uncle Eleazer's warm congratulations.[63]

During October 1871 Mrs. Sherman became so critically ill that Roger Sherman feared he might be unable to meet her medical expenses. He asked F. R. Lanier of Mississippi County, Arkansas, to redeem before due date the notes with which he had purchased

[61] *Ibid.,* December 1871—January 1872, February 13, 1872; James A. Garfield Papers (Library of Congress), J. H. Devereux to Garfield, April 19, 1872.

[62] Sherman Papers, *Pamphlets,* XVIII, No. 13.

[63] *Ibid.,* Sherman to Rev. K. Metcalf, April 10, 1871.

Pecan Point Plantation immediately before Sherman's marriage.[64] Mrs. Sherman recovered, and Roger Sherman prospered so well that in June 1872 he purchased a rambling frame house at 27 East Walnut Street, Titusville, a few steps from the Presbyterian church on Franklin Street. He hired a girl as help and in 1873 had a conservatory built.

By this time he had become a director of the conservatively managed Second National Bank, evidence of his interest in sound banking at the height of the great speculative boom. In banking as in railroad policy Sherman exhibited his independence of the Abbott associates.[65] His Pithole years had taught him a lesson that he did not forget.

[64] *Ibid.,* E. M. Conkling to Sherman, April 18, 1871, Sherman to F. R. Lanier, July 13, 1870.

[65] *Ibid.,* MS warranty deed, O. S. & Mary A. Harrington to R. Sherman, June 24, 1872, Sherman to W. W. Darne, August 6, 1872.

# CHAPTER II

# *The South Improvement Company Crisis*

ROGER SHERMAN'S sudden change in his railroad alignment from the Erie Ring to the Vanderbilts in mid-January 1872 occurred five weeks before the sudden disclosure in the *Titusville Daily Courier* of a secret alliance between the three great trunkline railroad systems and a mysterious South Improvement Company. The discovery was precipitated by a drastic increase in crude-oil freight rates. Rumor had it on February 20 that the South Improvement Company would enjoy such exclusive rebates from the trunklines as to enable it to monopolize the oil trade. Instead of restored railroad and oil-buyer competition for crude-oil shipments to New York City, which Sherman and Abbott were working separately to realize, the Oil Region was confronted with the prospect of having to sell to a single buyer.

This startling development threatened the position and prosperity of Titusville as the leading urban center of the Oil Region. It was proud of being the site of the first oil refinery, of being adjacent to Colonel Drake's well. Its Oil Exchange, of which John D. Archbold was Secretary, was the first of its type. It possessed a Board of Trade, six banks, two daily papers and the "independent, but not neutral" *Sunday Morning Press,* gas street lighting provided by the first fuel-gas pipeline in the world, the "Holly Water Works" under construction, and more than a mile of wooden pavement of the downtown "avenues" on "handsomely graded" Spring, Diamond, Franklin, Pine, and Washington Streets. The opera house was the best in northwest Pennsylvania, Colonel Henry declared. The "go ahead spirit" of nearly 10,000 inhabitants was characteristic of that era of widening frontiers.

The municipality dated from the charter of 1867. It possessed a mayor, common and select councils, school board, police force, and a fire department composed of volunteer companies of the type that had originated with Benjamin Franklin. Four public schools and a public park adjacent to the business section attested to the civic spirit of the municipality. A dozen churches represented "almost every profession of Christianity." Titusville then was predominantly Presbyterian, but was influenced among the enterprisers by a vigorous Episcopalianism led by Abbott, who was Senior Warden of St. James Church.

Although it had made the transition from borough to city only recently, Titusville had achieved a remarkably stable economy. Whereas other oil towns boomed and then declined quickly, its position as the urban center of the Oil Region rested upon its role as the wholesale distributor of provisions, lumber, machinery, equipment, merchandise, and coal to much of the area, upon its ability to finance an increasing volume of oil production, upon its ownership of refineries locally and on Oil Creek, and upon an increasingly diversified industry that was geared to the expanding needs of the petroleum industry. Engines, boilers, other parts of oil-well machinery, tanks for railroad oil tank cars, oil-storage tanks for wells and pipelines, oil-well tools, and beams for derricks were produced by the Titusville Manufacturing Company, Gibbs & Sterritt Manufacturing Company, Adam Good Brass Foundry and Machine Shop, Bryan, Dillingham & Company, and a planing mill. Roberts Brothers monopolized the manufacture of torpedoes. Thus, after Pithole and Pleasantville boomed and "broke," Titusville remained prosperous, confident in a future that to its inhabitants pointed toward metropolitan leadership in northwest Pennsylvania.

Some of the city's position rested upon its railroad connections. Passing through Titusville, the Oil Creek and Allegheny River Railroad connected at Oil City to the south with the Allegheny Valley Railroad and the Atlantic and Great Western Railroad and at Corry to the north with the Philadelphia and Erie railroads. The Franklin branch of the Lake Shore & Michigan Southern Railroad and the incomplete Jamestown and Warren Railroad provided indirect outlets. At Titusville was situated the headquarters of the Pennsylvania Tranportation Company, whose pipeline network was now challenged by the Union Pipeline Company and short local pipelines in new oil districts.

As early as 1869 Andrew Cone and Walter B. Johns in *Petrolia* declared that the city bade "fair to retain its position as one of the most thriving inland cities of our State," and that its citizens were "the most enterprising in the Oil Region." Successful producers built fine residences there for their families, spending weekdays in the oil districts drilling, cleaning, and torpedoing old wells, buying oil leases, and prospecting for oil with every device from divining rods to the geologists' handbooks. The more efficient and enterprising moved constantly into new districts, at times pioneering in their development. The producers' increasing wealth and confidence contributed largely to the optimism with which Titusville residents viewed the future. All elements joined in the effort to build the city in all aspects of economic, cultural, and social life simultaneously.

Most of Titusville lay north of the "flats" and railroad tracks along Oil Creek, where the refineries, oil-storage tanks, and certain warehouses were situated. Fifty to a hundred yards north of the railroad station was the business district, where the hotels, stores, banks, and office buildings were of brick. Extending up Spring and Franklin streets as well as parallel to Oil Creek, the solidity and prosperity of the area impressed visitors and set a standard that other oil towns could not match. To the north toward a wooded hill were situated blocks of residences with churches, schools, and occasional open spaces.[1]

[1] *Titusville Daily Courier,* October 1, 1870—February 1, 1872; *Titusville Morning Herald,* 1868–1872; Cone and Johns, *op. cit.,* pp. 569–570; Henry, *op. cit.,* pp. 589–590; "Petroleum, Chapter X," *Petroleum Age,* I (September 1882), 320; "Petroleum, Chapter XVII," *ibid.,* II (April 1883), 472; *Rules and Regulations of the Titusville Gas and Water Company* (revised, n.d., n.p.) in Sherman Papers, *Pamphlets,* XLIV, No. 10; Bates, *op. cit.,* pp. 304–312; Roland Harper Maybee, *Railroad Competition and the Oil Trade, 1855–1873* (Mt. Pleasant, 1940), *passim;* Allan Nevins, *The Emergence of Modern America, 1865–1878* (*A History of American Life,* VIII) (New York, 1928), pp. 40–41.

Titusville's ebulient spirit and growing wealth led it to dream of rivaling the great oil-refining centers of Cleveland, Pittsburgh, Philadelphia, and New York City. Each of these possessed larger capital, larger refineries, aggressive oil buyers, and railroads reaching into the Oil Region. Investments there enabled the rival eastern cities especially to control much of its oil-producing, transportation, and financial enterprises. Their metropolitan rivalry for leadership in the Oil Region was an unforgettable part of the background of the South Improvement Company.

This attempted coup threatened the profit position and very existence of the oil producers, oil buyers and shippers, the subsidiary industries, and oil-well suppliers as well as the banks of the Oil Region. The sudden 100% increase in crude-oil freight rates imposed on February 20 threatened to depress the price of oil drastically on the Oil Exchanges.[2] Strangely, the rate on refined oil from Cleveland or Pittsburgh to New York City remained at $2.00 a barrel, but rumors of large rebates upon this to be paid to the South Improvement Company appeared to doom the Oil Creek refineries which had been struggling to survive in the face of earlier discriminations favoring the Rockefeller group, certain Pittsburgh refiners, and Bostwick & Tilford. Caught unaware, the speculators on the Oil Exchanges at Titusville and Oil City had been unable to hedge against the strong bear movement that the discovery of the South Improvement Company's privileged position precipitated.

The threatened revolution in the oil trade presaged by early estimates of the prospects of that mysterious company threatened also to stop the accumulation of capital in the hands of local merchants, producers, manufacturers, and investors which had encouraged hopes of regional economic independence.[3] These hopes the *Titusville Morning Herald* and other newspapers had voiced when they boasted of the advantages of a refining industry situated adjacent to the wells.[4] Regionalism had emerged during 1871 and

[2] *Titusville Daily Courier,* February 20, 1872.

[3] "Capital in the Oil Regions," *Petroleum Age,* I (December 1881), 35.

[4] Allen Nevins, *A Study in Power: John D. Rockefeller, Industrialist and Philanthropist,* I (New York, 1953), 87.

1872 as the goal of civic leadership to replace the economic dependence upon the four rival metropolitan centers, a kind of dependence analogous to that of western mining centers upon Montgomery Street in San Francisco. Instead of a highly competitive, speculative petroleum industry controlled increasingly from within the Oil Region for its enrichment and development, the prospect presented by the South Improvement Company, unless quickly quashed, was the subjugation of the industry and the region alike to an all-powerful, outside monopoly. In an era distinguished by ruthless stock and gold-market coups of the Erie Ring the attempted South Improvement Company coup set a record in scope and in the amount of the damage it threatened to inflict upon legitimate industry and trade.

Close examination of its origin as revealed in contemporary sources and in Allan Nevins' *Study in Power* illumines the objects of the organizers and their sponsors and defines the formidable problem that the coup presented to the Oil Region. The South Improvement Company was the instrument of the especially railroad-favored group of oil refiners and oil buyers. It was the by-product of the bitter oil-rate war of 1871 among the railroads that was precipitated by the completion of the Lake Shore's branch to Franklin and its bid for control of the oil traffic to Cleveland.[5] While this encouraged oil shipments and sustained the price of crude oil in Titusville, it handicapped the Oil Creek refiners by diminishing further the transportation costs of the larger refiners of Cleveland and Pittsburgh. Simultaneously, the steady increase in the capacity of the refining industry to treble the daily oil production[6] had reduced the profit margin on refined oil. Hence, although the Rockefeller group in Cleveland undoubtedly made money in refining in 1871, it acquired a distaste for intensive competition similar to that which crystallized in the office of the Lake Shore whose managing director, Amasa Stone, had recently been allowed to

[5] Garfield Papers, Devereux to Garfield, April 19, 1872, which was not used by Williamson and Daum, *op. cit.*, pp. 346–350.

[6] *Titusville Morning Herald*, February 26, 1872; Stocking, *op. cit.*, pp. 12–13.

purchase a block of stock in the Standard Oil Company.[7] He too developed an antipathy to the competitive system because the oil-rate war seriously diminished hitherto exorbitant profits. The secure profit position of the Roberts Brothers' torpedo business and of the Pennsylvania Transportation Company's quasi-monopoly offered suggestive illustrations of an alternative policy.

Available sources indicate that the decisive, immediate impulse toward monopolistic organization of the petroleum industry came from the railroads. Heavy losses in oil freight revenue during the crude-oil rate war obliged the executives of the Pennsylvania and Erie trunkline systems which had "suffered equally" to offer to the Lake Shore-New York Central system "a compromise, which was agreed to." This was a plan to pool the oil traffic. P. H. Watson, manager of the Jamestown and Warren Railroad, a Vanderbilt subordinate, was appointed "umpire" to enforce the pool's pro rata provisions and the higher schedule of oil freight rates agreed upon. At this juncture certain Cleveland oil refiners, the Standard Oil and Clark, Payne & Company, joined leading Pittsburgh refiners in asking that "Mr. Watson should settle their differences" also. They wished to deal with the excess capacity of the refining industry in a manner that would restore prices on refined products and remove the threat of competition by Oil Creek refiners.

John D. Rockefeller had approached his leading rivals during the negotiation of the trunkline oil-traffic pool with a view to stabilization of the refining industry by mergers. "Cooperation," he called it. When approached by the leading refiners, Watson proposed that the Standard Oil, Clark, Payne & Company, and Bostwick & Tilford join the three refining companies especially favored by the Pennsylvania Railroad in subscribing to the stock of the South Improvement Company, whose charter had been secured at Harrisburg by the secretary of Thomas A. Scott, President of the Pennsylvania Railroad Company. When the leading executives of these firms bought the stock, Watson succeeded in combining the leading refiners and leading oil-buying company in a manner that facilitated enforcement of the carriers' oil-traffic pool. The lack of representa-

[7] Nevins, *Study in Power,* I, 90, 94–96.

tion of the Titusville and Oil Creek refiners in the South Improvement Company was significant. One thousand shares were taken by the executives of the leading refiners of Pittsburgh and Philadelphia, Lockhart, Frew & Company, Frew and Company, and the Atlantic Refining Company. Nine hundred shares were taken by five representatives of the Standard Oil, Clark, Payne & Company, and Bostwick & Tilford. Watson, President of the South Improvement Company, took the remaining 100. As the Vanderbilt ally of the Standard Oil he enabled the northern refiners and Bostwick & Tilford to exert within the South Improvement Company an authority equal to that of the favorites of the Pennsylvania Railroad. He was thus in a position to act as umpire of the projected refiners' pool in alliance with the trunkline oil-traffic pool that he administered.[8]

Led by Rockefeller the South Improvement Company then negotiated a secret rate contract with the three trunklines. This was signed by all parties on January 18, 1872, after Thomas A. Scott had objected in a manner that obliged the former to promise that the entire refining industry would be taken in. Scott predicted, however, that unless the oil producers were included also the plan would fail. The railroad rate contract provided that the open rate on crude oil from the Oil Region to Cleveland should be increased from 40¢ to 80¢ a barrel, and from the Oil Region to Philadelphia, New York, and Boston, to $2.41, $2.56, and $2.71, respectively. The rate on refined oil from Cleveland or Pittsburgh to New York remained at $2. Upon all crude-oil rates the South Improvement Company was to receive rebates ranging from 40%

[8] *Ibid.,* I, 100, 103–106, 109, 412, n. 19; Garfield Papers, Devereux to Garfield, April 19, 1872; *Titusville Morning Herald,* February 21, 26, 1872; *New York Tribune,* March 7, 9, 26, 1872; *History of the Rise and Fall of the South Improvement Company* (Titusville, 1872), pp. 30–41, 76–97; United States Industrial Commission, *Preliminary Report on Industrial Combinations,* I (Washington, D.C., 1900), 421–422, statement by Patrick C. Boyle, September 6, 1899; *The Derrick's Handbook of Petroleum* (Oil City, 1898), p. 168, quoting *Petroleum Centre Record,* February 20, 1872; Committee on Manufactures, *Investigation of Certain Trusts,* House of Representatives Report No. 3112 (Washington, 1889), p. 353.

to 50% and on all refined-oil rates from 25% to 45%, both upon its own shipments *and upon those of all others.*[9] When it is recalled that the five refining companies participating in the South Improvement Company manufactured less than 35% of the total production of the industry[10] it is evident that no more powerful weapon was ever placed in the hands of would-be monopolists by their railroad allies.

In return for these extraordinary rate concessions the South Improvement Company contracted with the Pennsylvania Railroad for itself, the Erie, and the New York Central systems, to allocate the oil traffic to the Atlantic coast between them in proportions of 45%, 27.5%, and 27.5%, respectively. The terms of the contract made it clear that those carriers expected that the South Improvement Company would quickly engross the entire oil trade so as to fulfill this provision. To facilitate this "evening" of the oil traffic the railroads contracted to inform the South Improvement Company in full of all oil shipments by its competitors and also to vary the official oil freight rates upon its demand in any manner necessary to enable it to overcome their competition! With five years' life, terminable only after a year's notice, this contract was clearly intended to produce an inclusive monopoly of oil shipping on the three great trunklines and their subsidiaries.[11]

No sooner had this unprecedented document been signed than Rockefeller proceeded to utilize the decisive rate advantage that it granted the Standard Oil to urge leading refining competitors in New York, Cleveland, and the Oil Region to sell out to the Standard Oil on its terms (its appraised value of their works plus oil on

[9] Ida M. Tarbell, *History of the Standard Oil Company,* I (New York, 1904) 56–59, 281–285; [Roger Sherman], *An Appeal to the Executive of Pennsylvania. An Address to Gov. John F. Hartranft, invoking the aid of the State against the unlawful acts of corporations, Presented August 15, 1878* (Titusville, 1878), pp. 5–6, 12–19; Nevins, *Study in Power,* I, 105–107, 109–110; Hans B. Thorelli, *The Federal Antitrust Policy: Origination of an American Tradition* (Baltimore, 1955), pp. 92–93; *History of the Rise and Fall of the South Improvement Company,* pp. 5–6, 76–96.

[10] Cf. Stocking, *op. cit.,* pp. 12–13; Tarbell, *op. cit.,* I, 68 n.

[11] Tarbell, *op. cit.,* I, 281–288.

hand). Late in 1871 the Standard Oil had purchased Bostwick & Tilford secretly.[12] Now it absorbed Clark, Payne & Company,[13] which gave the Rockefeller coterie all of the northern-held shares of the South Improvement Company except those owned by Watson. Before the premature increase of oil freight rates by a subordinate railroad official in the Oil Region precipitated disclosure of the conspiracy on February 20, 1872, the Standard Oil's executives thus controlled 45% of the stock, the largest single holding. Then, during the panic in Cleveland among its refining competitors that was precipitated by the disclosure, they acquired twenty-one out of twenty-three of them.[14] In negotiations for the acquisition of other competitors elsewhere and in overtures to leading oil producers to come into the South Improvement Company the Standard Oil executives acted as if their company were the instrument of the railroad pool. Irrespective of whether or not it expected to collect the rebates promised the South Improvement Company the Standard Oil possessed its own contracts for large drawbacks and rebates with the Erie and New York Central systems.

Although representatives of the Oil Region did not learn the terms of the South Improvement Company contract with the railroads until a month after the increase in oil freight rates, Rockefeller and his lieutenants apparently divulged enough of them confidentially to companies that they approached with a view to their purchase for this to leak out and circulate widely.[15]

As far as Roger Sherman was concerned, he was affected by the great conspiracy in terms of its impact upon Titusville and the Oil Region. While not at that time engaged in oil production so far as surviving records disclose, his father-in-law at Pleasantville was probably a producer. Many of Sherman's clients were producers. Others were businessmen and industrialists in Titusville and along Oil Creek. His law practice and future were identified with them. As

[12] Nevins, *Study in Power,* I, 134; Johnson, *op. cit.,* p. 29, erroneously dates this early in 1872.

[13] Nevins, *Study in Power,* I, 133–134.

[14] *Ibid.,* I, 135–151; Tarbell, *op. cit.,* I, 63–68; *Derrick's Handbook,* p. 177; *New York Tribune,* March 8, 1872.

[15] Nevins, *Study in Power,* I, 95–120.

a friend of N. M. Allen, who was a large creditor of the *Courier,*[16] and as a contributor to its columns, Sherman was fully aware of the scope of the danger that the newly disclosed South Improvement Company presented to Titusville and the Oil Region. He knew that the railroad rate favors the South Improvement Company was rumored to enjoy had some precedents in the Erie Ring's earlier favoritism to privileged shippers. As the attorney for such prominent oil producers as Patterson, the Emery brothers, the Fertigs, and Dr. G. S. Shamburg he could appreciate the impact of the drastic increase in oil freight rates upon their prospects. As Secretary of the Titusville Publishing Association he occupied a unique position in which he could assist the *Courier* to render a public service by publishing as complete an exposure as was possible of the South Improvement Company coup.

The *Courier's* exposure on February 20 of the existence and character of the South Improvement Company, assisted by stories in the *Morning Herald* and *Oil City Derrick,* precipitated a popular explosion. Basing its comment upon data supplied by A. P. Bennett and John D. Archbold, who had made a futile fight in behalf of the Oil Creek refiners in New York against a rumored increase in freight rates, the *Courier* declared that the South Improvement Company "looks like a vast conspiracy, organized to seize upon and appropriate to its members such tremendous advantages, as shall enrich them at the expense of the great mass of oil refiners, dealers, brokers, together with the entire producing interest of the country. From all we can learn it is contemplated to make good the threat to 'dry up Titusville.' The giving up of the W. & V. R. R. we believe to be included in the program." The *Courier* appealed to Titusville and the Oil Creek producers and refiners to confront this danger, prepare for the worst, and to act before it was too late.

On the morning after the *Courier's* editorial appeal virtually all business was suspended in Titusville. The great conspiracy became "the all-exciting topic of conversation." The *Courier* then announced that the Oil Creek refiners were not included in "the

[16] Bates, *op. cit.,* pp. 317–318.

gigantic combination."[17] It then drew upon all available talent, obviously including Roger Sherman, in a vigorous campaign against the South Improvement Company. The *Courier* rivaled the *Morning Herald* and the *Oil City Derrick* in attacking it,[18] and in rallying the Oil Region to oppose it.

The public responded in an unprecedented manner. The outraged fury of all elements in Oildom's business community was expressed in mass meetings, fiery editorials of the press, resolutions of the Oil Exchanges, the organization of the producers into the Petroleum Producers Unions, and the appointment of committees by producers, refiners, and mass meetings to secure from trunkline executives the actual facts and to fight the sudden, ruinously large oil freight rate increases. Despite its competitive, at times acrimonious, pluralism the Oil Region was welded suddenly by the great danger into an angry, determined unit.[19]

Other than by unsigned contributions to the newspaper and his behind-the-scenes assistance in planning the *Courier*'s campaign, Sherman did not figure prominently in the hastily improvised opposition of the Oil Region to the great conspiracy. His public standing, evidenced by his recent leadership on behalf of the Warren & Venango Railroad, was not affected when Jay Gould's telegram was read to the huge mass meeting in Paschal Opera House on the night of February 27.[20] The telegram, admitting that Gould and the receiver of the Atlantic and Great Western Railroad, and executives of the Lake Shore, New York Central, and Pennsylvania railroads had signed a rate contract with the South Improvement Company, validated the hasty deductions that had been drawn heatedly after the premature increase in the crude-oil rate from Warren to New York City from 87¢ to $2.14 had been put into effect on the

[17] February 21, 1872.

[18] *Titusville Daily Courier,* February 24, 28, 1872; *Titusville Morning Herald,* February 21, 22, 24, 26, 27, 1872.

[19] *Pittsburgh Leader,* February 29, 1872, quoted in *Titusville Morning Herald,* March 2, 1872; Harold M. Helfman, "Twenty-Nine Hectic Days: Public Opinion and the Oil War of 1872," *Pennsylvania History,* XVII (April 1950), 121–138.

[20] *Titusville Morning Herald,* February 28, 1872.

Jamestown and Franklin branch of the Lake Shore on February 19.[21] Professional obligations to his clients prevented Sherman from publicizing what he knew of how the Erie Ring and the Pennsylvania Transportation Company had built up the Standard Oil and other Cleveland refiners by means of secret exclusive freight-rate drawbacks. Had he done so he might have become a prominent leader in the resistance to the South Improvement Company such as John D. Archbold of Titusville and S. C. T. Dodd of Venango County were becoming.

The Petroleum Producers Unions, whose sixteen districts comprised the entire Oil Region, sent a Committee of Ten in cooperation with the Oil Creek refiners to New York City. Their joint committee interviewed the trunkline presidents in an attempt to secure a modification of the new oil freight-rate schedule. Elisha G. Patterson, Sherman's client and former President of the Petroleum Producers Association, was a prominent member of a delegation that was sent to Harrisburg to secure repeal of the South Improvement Company's charter and to lobby for a free-pipeline bill that would enable the producers to smash the monopoly of the Pennsylvania Transportation Company in which Gould was so influential.[22] Lewis Emery, Jr., another of Sherman's clients, was an influential leader of the Producers Unions. Perhaps Sherman was consulted also by Henry Harley and George K. Anderson before they went to New York City. There they persuaded the trunkline presidents to declare in favor of granting to all oil shippers the rates that had been granted exclusively to the South Improvement Company. This was the first breach in the powerful phalanx of the conspiracy.[23] An ironclad boycott was promptly clamped upon the South Improvement Company and its affiliates by the oil buyers and oil shippers. This the producers extended to all railroad oil shipments

[21] The official increase went into effect on February 26. See *Titusville Morning Herald,* February 27, 1872; *Derrick's Handbook,* pp. 168–169. Tarbell, *op. cit.,* I, 71, gives February 22 as the date.

[22] *Titusville Daily Courier,* October 14, 1870, March 25, 1872; Helfman, *op. cit.,* 130–137; Tarbell, *op. cit.,* I, 72.

[23] *Titusville Morning Herald,* March 20, 1872.

so as to oblige the trunklines to cancel their contract with that hated company formally.[24]

At Titusville the Oil Exchange led the fight against the great combination. It joined the Producers Unions in many measures. Among these was a demand that Pennsylvania's United States Senators instigate a congressional investigation of the conspiracy and press for the adoption of a policy of national railroad regulation as the most effective method of removing freight and service discriminations.[25]

All members of the Oil Exchanges, all producers, and all refiners recognized, as the *Pittsburgh Leader* observed on February 29, that the success of the combination would create "but one buyer of oil in the whole oil region" with power to "fix prices just to suit itself" and to control "the whole oil production." For thousands of men of property in the Oil Region and for its numerous laborers and clerks this would bring "ruin and distress."

On March 23 Sherman joined the Titusville Oil Exchange as it pledged its members to boycott the South Improvement Company and its members.[26]

When the *Morning Herald* proposed that an "Oil Producers Railroad" be constructed to Buffalo to break the trunklines' hold on the oil traffic and to protect the Oil Region from future monopolies and railroad combinations, Sherman was moderately interested. Ten days later, on the day after the three trunklines had "cancelled" their rate contract with the South Improvement Company he subscribed $250 to the stock of the projected railroad.[27] Perhaps his commitment to the Warren & Venango Railroad had prevented him from subscribing earlier. Eventually the new road was completed as the Buffalo, Titusville, and Pittsburgh Railroad.

All elements in the Oil Region recognized now that rate discrimination practiced by the railroads was the chief cause of their prob-

[24] *Ibid.*, March 7, 1872; Helfman, *op. cit.*, p. 137; Johnson, *op. cit.*, p. 20.

[25] Helfman, *op. cit.*, 135–136.

[26] *Titusville Morning Herald*, March 25, 1872; Bates, *op. cit.*, p. 315.

[27] *Titusville Morning Herald*, March 22, 29, 1872.

lems. Elimination of discriminatory procedures became the major object of the protesting groups and their leaders. As they analyzed the problem they recognized that carrier-controlled pipelines, such as the Pennsylvania Transportation Company and the Union Pipeline Company now controlled by the Empire Transportation Company, the Pennsylvania Railroad's subsidiary, entered into the situation. This explained why the producers and refiners demanded a state free-pipeline law and national railroad regulation simultaneously. An especially interesting feature of those "Twenty-Nine Hectic Days," as Harold M. Helfman [28] terms the interval between the discovery of the great conspiracy and the repeal of the South Improvement Company's charter, was the mobilization of almost unanimous hostility to the "omniverous monopoly." For years afterward a dynamic antimonopolism would characterize public opinion in the Oil Region.

A number of developments contributed to the defeat of the great combination. Among these was the proposal for an oil producers' railroad and the promise of Pennsylvania's United States Senators to support the Oil Region against the "outrageous railroad monopoly." Harley's intervention with the trunkline presidents helped. More important were the adamant boycotts of the South Improvement Company and its affiliated firms and of the trunklines by oil shippers outside the combination. During negotiations with the trunkline executives the producers' and refiners' committee firmly demanded equitable treatment of the Oil Region and its oil men. The Producers Unions as a powerful trade association had to be conciliated. The New York refiners supported the joint committee which induced them and the Oil Creek refiners to decline to sell out to the Standard Oil. Gould's overthrow and ouster from the presidency of the Erie Railroad at that company's annual stockholders' meeting on March 11 [29] facilitated the revocation of the trunklines' contracts with the South Improvement Company and helped to secure the trunklines' promise of "equal and open rates for all."

[28] Helfman, *op. cit.*

[29] *New York Tribune,* March 12, 1872.

The enactment by the Pennsylvania legislature,[30] despite Thomas A. Scott's opposition, of a free-pipeline law for the five counties of the Oil Region also helped.

As a behind-the-scenes adviser of the *Courier,* producers, pipeline operators, and businessmen, Sherman was in a position to exert his influence constructively. He was undoubtedly interested, as were the producers and refiners, in the greatly enhanced importance and aggressive tactics of the Standard Oil Company of Ohio. The Oil Region's antipathy to the South Improvement Company was quickly extended to the Rockefeller firm, which was regarded then and thereafter as its alter ego. Its attempt to persuade important producers to break the boycott on oil shipments in its behalf was much resented.[31] An interview in the New York press with Henry C. Ohlen, the prominent New York oil buyer who supplied Peter Lombard, the refiner, when republished in the Oil Region's newspapers, divulged the Erie Ring's and Rockefeller's importance in the pattern of freight rate discriminations and South Improvement Company activity.[32]

This was confirmed and amplified later by the hearings in April of the congressional investigation of the conspiracy, the findings of which appeared seriatim in such journals as the *Morning Herald.*[33] These made it evident that Rockefeller had played a major role in the great conspiracy. As he maneuvered thereafter to develop more effective methods of achieving the monopoly of oil refining, his foes of the spectacular fight of the Oil Region against the South Improvement Company would attribute to him a stubborn adherence to its great objective and methods and a remorselessness which popular stereotypes have ever attributed to the monopolist.

Sherman was able to detect Rockefeller's equivocation when the latter denied on April 8 that the Standard Oil held any contracts with the railroads. Sherman knew that its contract with the Erie

[30] *Titusville Morning Herald,* March 19, 1872.

[31] *Ibid.,* March 25, 26, 29, 1872.

[32] *New York Herald,* March 19, 1872, in *Titusville Morning Herald,* March 19, 1872.

[33] *Titusville Morning Herald,* March 29, April 6, 22, 1872; *History of the Rise and Fall of the South Improvement Company,* pp. 33–57, 68–96.

Railroad for drawbacks on crude-oil shipments to Cleveland would not expire until 1873. Subsequent investigation would reveal that the Standard Oil had signed recently a new secret rebate contract with the New York Central. Interestingly, William H. Vanderbilt denied the existence of this new contract on April 9 in a telegram to the Oil City meeting of the Producers Unions. Since Sherman and his friends were now cognizant of the previously existing trunkline policy of favoring large oil refiners and shippers especially, they probably doubted the truthfulness of both statements.

However, they joined in the Oil Region's celebration of victory at the overthrow of the South Improvement Company when its rate contract was revoked on March 25,[34] and again when its charter was revoked on April 2.[35] The permanence of a railroad-controlled pipeline monopoly within the Oil Region appeared to be impossible after enactment of the new free-pipeline law. Although the fruits were shorter-lived than the allied producers and independent refiners anticipated in the spring of 1872, they had won a notable victory. In the first great conflict between free enterprise and the new spirit of aggressive, ruthless, industrial monopoly, the free enterprisers had won. Although Roger Sherman's full contribution to this victory cannot now be ascertained, it is evident that from that conflict he learned lessons that he never forgot. These contributed importantly to the values, antimonopolistic position, and antimonopolistic tactics that featured his subsequent career as a champion of the Oil Region and his efforts to protect its business interests against monopolistic dominance.

[34] *Titusville Morning Herald,* March 25, 1872.
[35] *Derrick's Handbook,* p. 176.

# CHAPTER III

# *Achievement of Professional Stature*

THE oil war in northwest Pennsylvania during February to April 1872 symbolized the short-lived popular revolt of that year against the evils of "The Gilded Age." So Mark Twain termed those postwar years in his fictional portrayal of the promoter-speculators and the corrupt political scene. The oil war was fiercely antimonopolist, double-edged in its fight against industrial and railroad monopoly. The leaders learned of the ramifications of railroad monopoly elsewhere from the *Titusville Morning Herald,* which told how the anthracite coal carriers had gained control of the anthracite coal industry. The demand of the *New York Tribune* for legislative railroad regulation expressed a metropolitan viewpoint that would ultimately produce the Hepburn Committee investigation into the railroads of New York in 1879.[1]

Appreciation of the magnitude of the victory over the South Improvement Company, which had been accompanied by trunkline assurances to the joint producers' and refiners' committee of equal treatment to all oil shippers, gave the Oil Region and its leaders a sense of power. This made some overconfident; others were more

[1] *New York Tribune,* March 8, 1872; *Titusville Morning Herald,* February 27, 1872.

realistic. The *Morning Herald* warned that the northern branch of the South Improvement Company, the Standard Oil, would survive unless crushed also. This company, it prophesied, would receive similar secret railroad favors—as it actually had from the Vanderbilts.[2] After the boycott of the railroads was discontinued, that of the Standard Oil continued. The Standard Oil defeated this boycott by inducing two producers to sell to it and because William H. Abbott successfully opposed continuance by the Titusville Oil Exchange of the boycott of the participants in the South Improvement Company.[3]

The Rockefeller company and Bostwick & Tilford, its subsidiary, resumed purchases on a greatly enlarged scale, since the Standard Oil had emerged from the oil war in possession of more than thirty-four per cent of the total oil-refining capacity of the industry.[4] Although this was insufficient to enable it to establish its price leadership in manufactured oil products, Rockefeller was able with the assistance of former associates in the South Improvement Company to persuade many of the surviving independent refiners to join in a Refiners Association in August 1872.[5]

This so alarmed the producers that they rallied again around William Hasson, who had led them in the oil war, to organize the Producers Protective Association. Fearing that once more they might be confronted with a single buyer of crude oil the Association's members attempted to create a unified, cooperative storage and marketing agency and to supervise the curtailment of production in order to achieve a maximum price for their product. These measures, they believed, would free them from the manipulations of "rings and combinations of speculators" on the exchanges. Hasson sought to advance the interests of the petroleum trade by establish-

[2] *Titusville Morning Herald,* March 22, 25, 1872; *New York Tribune,* March 26, 1872; Tarbell, *op. cit.,* I, 100.

[3] *Titusville Morning Herald,* April 3, 6, 9, 10, 1872; *Derrick's Handbook,* p. 176.

[4] *Titusville Morning Herald,* March 14, 26, May 18, 1872; Nevins, *Study in Power,* I, 175, 178; Tarbell, *op. cit.,* I, 68.

[5] *Derrick's Handbook,* pp. 180, 185; Johnson, *op. cit.,* p. 31; Tarbell, *op. cit.,* I, 106–109.

ing uniform standards of measure, value, and business methods, and "if necessary" by providing the "refining and marketing at cost" of the crude oil so as to "break down all possibility of combination against the producing interest." [6] Roger Sherman was to be identified throughout his career with efforts to effect realization of the latter policy.

Hasson's program included cultural objectives—not only good homes for producers and oil-well workers but also good schools, libraries, lecture courses, art galleries, and "entertainments of a higher order instead of the low houses of dissipation and the cheap amusements which attend the unsettled condition of the region—and good churches." Obviously, a cooperatively organized producers' monopoly of crude oil could not successfully abolish monopoly in the Oil Region. The rival combinations soon collapsed.[7]

Producer antimonopolism was expressed also by continued resistance to the Roberts Brothers' torpedo patent. Elisha G. Patterson, Sherman's client, was chairman of the general committee that conducted the fight against this in the courts.[8]

Soon the producers would find another monopolistic foe in the gathering pipelines as they were concentrated under the control of a few large corporations. The pipelines continued to be intimately related to trunkline competition for the oil traffic. After a two years' renewal of this competition the trunkline executives returned to pooling in another attempt to restore and stabilize oil freight rates. Negotiations were completed in New York City on September 9, 1874. As announced by the notorious Rutter circular the pool included the pipelines, which Henry Harley had urged, so as to prevent pipeline and carrier competitive rate cutting. Like the South Improvement Company conspiracy of 1872 the circular penalized the refineries situated in the Oil Region. Those that received their crude oil by pipeline were deprived of the pipage railroads paid to

[6] *Address and Constitution of the Petroleum Producers Association* (Oil City, 1872), pp. 1–5, 8, 11–12; Johnson, *op. cit.*, pp. 31–32; *Derrick's Handbook*, p. 186; Tarbell, *op. cit.*, I, 110–125.

[7] *Address and Constitution of the Producers Association*, pp. 5–6.

[8] *Titusville Daily Courier*, February 19, 1872.

pipelines. Since by this time the Standard Oil had become active in the pipeline business in competition with the Pennsylvania Transportation Company and the Union Pipeline Company, it became a party to the trunkline oil-traffic pool and a beneficiary of its favors. The Rutter circular's discrimination against the Oil Region provoked explosive opposition and immediate organization there against this new combination.[9]

Since Roger Sherman's law practice at Titusville was intimately related to the petroleum industry he reacted to these developments as did other men who were unassociated with the railroads and the Standard Oil. Some years intervened, however, before he was called upon to play a leading role in the fluctuating contest between the Oil Region and the externally controlled corporations which were attempting to subordinate the area to themselves.

He was drawn into the situation first as the result of publicity exposing the pipeline companies' and the Oil Ring's speculative "bearing" of oil on the exchanges and also the pipelines' failure to report oil stocks in storage. In order to curtail their speculation in oil and require better performance of their common-carrier functions Sherman drafted a bill early in 1874. This required monthly pipeline and quarterly railroad reports of oil shipped, of oil stored by the former, and of the oil necessary to cover outstanding company issued pipeline certificates certifying to oil in storage. This bill was enacted on May 15, 1874.[10]

However, the Wallace Corporation Act adopted a month earlier had partially invalidated the free-pipeline law of 1872. Thus, while Sherman's pipeline regulatory act had yet to be proven effective the situation appeared to favor pooling or combining existing pipelines in a manner adverse to the producers' interests.[11] The Wallace Act also provided an apparent basis for the trunkline-pipeline pool announced by the Rutter circular. Simultaneously, Sherman and his friends were confronted by the refusal of the pipeline companies to

[9] *Investigation of Certain Trusts,* p. 363; Johnson, *op. cit.,* pp. 28–29, 37–38; Tarbell, *op. cit.,* I, 136–143.

[10] After the quarterly railroad reports of oil shipments section was deleted, Sherman, *op. cit.,* p. 71; Johnson, *op. cit.,* pp. 32–33.

[11] Johnson, *op. cit.,* p. 34.

conform to the regulatory act of May 15. For a year they patiently gathered evidence of abuses.

In September 1875 Sherman drafted a sixteen-page statement for Elisha G. Patterson detailing the ways in which the regulatory act provided remedies for the multifarious abuses that Patterson had reported. The act had attempted to eliminate the fraudulent over-issuing of oil certificates. Violations of its provisions were punishable by fines up to $1,000 and imprisonment up to a year. After careful study Sherman informed Patterson that this regulatory act was constitutional and could be enforced. He insisted that evidence be gathered diligently in advance of prosecutions.[12]

As attorney for the producers, Sherman instituted actions three months later against a number of local and regional pipeline companies. These included civil damage suits and also criminal prosecutions against company officers based upon information supplied to public prosecutors in the counties where violations had occurred. In these suits Patterson appeared in the novel role of "prosecutor," having cooperated in gathering evidence and supplying it to the state attorneys. Sherman insisted that the bills of indictment be amply detailed before presentation to grand juries.

Patterson was ribbed broadly by the *Morning Herald* because of his unaccustomed role, although the editor conceded his consistency as an antimonopolist and the existence of the abuses that justified the prosecutions. Patterson, the editor asserted, was backed by capitalists and producers "who don't show up their hands," and remarked that he had retained as his counsel Sherman, Charles Taylor (Warren County's leading criminal lawyer), and a prominent Pittsburgh attorney. Sherman, the *Morning Herald* editor remarked, was "one of the longest headed of the practitioners of the Crawford County bar." [13]

Sherman entertained no illusions as to the integrity or solvency of the pipeline companies. To Dun & Company that same month he reported that the Pennsylvania Transportation Company probably had sufficient oil certificates in circulation "to swamp them if presented at once, . . . Those who have to deal with pipe lines should

[12] Sherman Papers, Sherman to Patterson, September 7, 1875.
[13] January 27, 1876.

be careful as they have unlimited facilities to swindle if they have the disposition." [14] This knowledge had also motivated his initiation of the pipeline regulatory act.

Guided by Sherman and his fellow counsel, Patterson sued for the maximum penalty for each violation in behalf of himself and the local school districts in Armstrong and Franklin counties. He instituted perjury prosecutions where company officers deliberately misstated the facts in their reports. As prosecutor he secured indictments and pressed trials of such company officials as Captain J. J. Vandergrift of the United Pipe Lines Company who had recently admitted the Standard Oil to part ownership. So harrassed, the pipeline company executives sued for peace.

Through their counsel the United Pipe Lines Company, Milton Pipe Line Company, Franklin Pipe Line Company, Sandy Creek Pipe Line Company, American Transfer Company, and Keystone Pipe Company formally complied with the law. They paid Patterson $7,000 for his "trouble expense & costs" incurred in instituting the damage suits against them. Thereafter they complied with the Sherman form of monthly report after having agreed "to make reports and conduct the business under the law." Since the United Pipe Lines and American Transfer companies were Standard Oil subsidiaries, Roger Sherman could conclude that his partnership with Patterson had emerged victor from its first bout with that giant organization.[15]

Since the Rockefeller group was now the *bête noir* of the produc-

[14] Sherman Papers, Sherman to Joshiah Marshall, January 8, 1876.

[15] *Ibid.,* Sherman to J. V. Painter, March 24, April 10, 1876, Sherman to E. G. Patterson, July 15, 1876, Sherman to J. McKessick, August 18, 1876, Sherman to J. Pitcairn, Jr., September 25, 26, 1876, Pitcairn, Jr., to Sherman, September 26, 28, 1876, MS Agreement between Patterson and Pitcairn, Jr., September 1, 1876, MSS Informations, Complaints, Indictments, and statements of penalties due in the suits initiated, MSS "Pipe Line Memoranda for Trial (copy)," January 10, 1876, official statement to pipelines of the information required monthly, August 1876, Venango County Grand Jury presentment against H. B. Van Velsor, agent of Sandy Pipe Line, MS September 1, 1876, agreement between E. G. Patterson and John Pitcairn, Jr., MS September 1, 1876, statement by E. G. Patterson. The form of monthly report finally agreed to was the result of consultation between Pitcairn, Jr., and Sherman.

ers, independent refiners, and general business interests of the Oil Region, this victory was not without its increment for Sherman's professional prestige. The financial returns of the litigation were miniscule, however. His share of the $1,000 allotted to the three counsel was insufficient to compensate him for his labor. Patterson's net share, $2,500, was also nominal when the magnitude of the interests involved and the time he spent in the litigation were considered. He and Sherman had performed an important public service that was highly beneficial to the Oil Region. After winning the major battle Sherman remained vigilant. In January 1877 he threatened to institute prosecutions to force some pipelines to comply with the regulatory act.[16]

By this time the Standard Oil-controlled United Pipe Lines had acquired most of the local and competing companies except the Empire Pipeline Company and the decadent Pennsylvania Transportation Company. Hence Sherman's and Patterson's victory offset only partly the damage inflicted upon the oil producers and the Oil Region by the Pennsylvania Railroad's success in preventing the re-enactment of the free-pipeline law that Lewis Emery, Jr., another of Sherman's clients, advocated in the Pennsylvania legislature during 1874 to 1883.[17]

The antimonopolist views that impelled Sherman and Patterson in this contest had already found classic expression before they launched their prosecutions. In May 1875 at a producers' meeting in Mendenhall's barn at Edenburg, Clarion County, Colonel J. A. Vera had compared the developing oil monopoly to the octopus. It grasped the producers in its pipeline feelers that reached out in all directions, he said, while its controlled railroads were "the suckers drawing them in and the Standard Oil Company at Cleveland is the devil fish that swallows them, body and soul!" As "The Octopus" it was regarded thereafter by the producers, oil buyers, oil shippers, and independent refiners of the Oil Region.[18]

[16] Sherman Papers, MS Agreement, Patterson and Pitcairn, Jr., September 1, 1876, Sherman to F. A. Morrison, January 26, 1877.

[17] United States Industrial Commission, *Preliminary Report,* I (testimony of Lewis Emery, Jr.), 102.

[18] *Petroleum Age,* VII (December 1887), 1797.

Undoubtedly with Sherman's assistance, Patterson in 1876 prepared a national antidiscrimination railroad bill and persuaded the General Council of the Petroleum Producers Unions to endorse it. The independent refiners of the Oil Region and a thousand Pittsburgh companies supported the bill when it was introduced by Congressman James H. Hopkins. Patterson testified in its behalf before the Committee on Commerce in an attempt "to show up the Cleveland monopoly that is squeezing the life out of the refineries at Titusville." So the *Morning Herald* reported, together with news that General J. H. Reagan of Texas, a committee member, was "hot for the investigation" of the Rockefeller combination. Congressman Reagan was the former Postmaster General of the Confederacy. In this manner Sherman and he contributed ex-Confederate support at an early hour to the rising movement for national railroad regulation.

The Oil Region and Pittsburgh were sanguine in their anticipation of the bill's immediate enactment. It did not pass the national House of Representatives before 1879, by which time the redrafted bill bore General Reagan's name. Sherman always alluded to this measure, which in altered form became the Interstate Commerce Act, with pride. Since he did not claim authorship it may be concluded that he had given Patterson advice privately regarding the contents of the original bill. The Interstate Commerce Act, he asserted later, originated in Titusville. The *Morning Herald* agreed. Sherman was among "the leading spirits in the movement" for its adoption, and the historian of Crawford County recorded in 1897 that he did "much to push vigorously the matter from time to time." [19]

These were the years in which Sherman achieved professional eminence. From 1872 to 1874 he was kept busy with his practice, although he devoted time to peripheral activities, some of which were related to the Oil Region's attempts to secure a continuance of railroad and market competition.

[19] Bates, *op. cit.,* pp. 800–801; *Titusville Morning Herald,* April 13, 22, 24, May 31, 1876, January 13, 31, 1879, May 16, 1889; Sherman, *op. cit.,* pp. 71–72.

His extraprofessional activities were various. In the spring and summer of 1872 he attended the Titusville Oil Exchange and the stockholders' meeting of the Oil Producers' Railroad Company. He continued as a director of the Second National Bank. In 1873 he became a director of the Titusville Printing Press Association and assisted in transfer of control and editorship of the *Titusville Daily Courier* to N. M. Allen. At the Second National Bank Sherman's friend, Cashier C. E. Hyde, was Secretary of the Board. That bank, interestingly, had settled in its new and unique building at the corner of Spring and Washington Streets. There it occupied the first of three floors of that limestone structure, whose numerous round arched windows were surmounted with raised eyebrows in high relief.[20]

During the autumn before the Panic of 1873 Sherman amused himself with writing for the *Courier* a series on the "Moravia Experience with the Spirits." This interested his aged uncle Eleazer Conkling, who thought the articles fair but written with great ability. He observed to Sherman that "you are not enough of an enthusiast not to look continually on all sides." Eleazer Conkling was thankful, erroneously, that his nephew was "not mixed up in politics." He advised him when another civil war occurred, "this is sure to come ere many years, to flee to some peaceful land." Reporting that a considerable number of Liberal Republicans and some Democrats were supporting Grant who was sure to win, he asked Sherman that November of 1872: "Can you swallow Greeley?"[21]

Sherman was drawn into the complex community of Titusville in other ways. In July he had been elected to membership in the volunteer Courier Hose Company. He participated in its October parade. He became a trustee of the Titusville Branch, Life Association of America. In January 1873 he helped to organize the Philomathic Union, a literary and debating society that scheduled discussions of "scientific, social, religious and political questions"

[20] Sherman Papers, C. C. Daffied to Sherman, August 27, 1873, D. W. Lockhard to Sherman, July 19, 1872.

[21] *Ibid.*, E. M. Conkling to Sherman, November 7, 1872.

and heard essays "on interesting subjects." He became a director of the Union, which also sponsored public programs in the Opera House.[22]

M. C. Beebe, Sherman's law partner, was sent by Crawford County as a delegate to the state Constitutional Convention of 1873. There he was placed on the railroad committee from which the Oil Region hoped for substantial reforms. He wrote Sherman regularly from the Convention, enlightening him upon the corrupt politics that prevailed in the Keystone State under the leadership of United States Senator Don Cameron. Beebe was skeptical whether or not "our work will ever amount to a s—" because the "ring at Harrisburg and the Politicians of Philadelphia of both parties are laying their plans to defeat our work and as fast as we do any work in the Railroad Committee the Rail Roads go to Harrisburgh and get a bill framed asking for all the legislation they want to cover our ground and get it in two days with out a call of the ayes & nays." He then reported to Sherman:

> The Convention are making such fearful exposures of the election frauds and corruption of the Legislature that they are coming down for us like a "Thousand of Brick" or Ten Thousand rather so the prospect is gloomy for the future for the whole body politic the people have become fearfully demoralized but still my only hope is that the people themselves will see it and [reform] I mean all parties for Republicans can buy Democrats like sheep for the shambles and at the same price. . . . This is all *sub Rosa.*[23]

As a representative in the next legislature Beebe introduced Sherman's and his school bill and persuaded the governor to sign it. However, because of the "Reform influence" he was unable to secure all the private legislation that their law firm desired. He was not above employing what influence he could muster in behalf of

[22] *Ibid.,* Titusville Citizens Corps "Invitation," November 11, 1872, F. W. Broughton to Sherman, July 11, 1872, Courier Hose Company "Invitation," October 3, 1872, *Pamphlets,* XVIII, No. 13, *Charter, Constitution and By-Laws of the Titusville Citizens Corps of Titusville, Crawford County, Pa.* (Titusville, Pa., 1873).

[23] Sherman Papers, M. C. Beebe to Sherman, n.d.

this, however. Such, for example, was a general bill based upon New York precedent that he induced Sherman to draft that would have helped them in the complicated "Harsh case" of a lady client involving the validity of a land title. However, the opposition of S. C. T. Dodd in the House and the indifference of the chairman of the Senate Judiciary Committee blocked action on this.[24]

Thereafter Sherman focused a sharp eye upon the legislature during its annual sessions at Harrisburg. In 1874 Frederick Bates received his solicited improvements for a bill on "private corporations" and consulted him on a similar bill passed by the Senate. During that same season Congressman James Shakely consulted him regarding the provisions of a House bill to reorganize the judiciary. Sherman read it carefully, recommended changes so as to reduce further the number of appeals that could be made to the United States Supreme Court, and approved the bill because it would contribute substantially to this end. Drawing upon Shakely's letter he wrote a column, "Washington Correspondence," for the *Courier* in which he explained the Congressman's position.[25]

Sherman's high reputation in the Pennsylvania bar and his influence on legislators became apparent in January 1877, when State Senator C. W. Stone consulted him on the proposed repeal of an Act of 1871 that regulated opinions of the Pennsylvania Supreme Court so as "to restore the old practice." If Sherman agreed, Stone requested that he write "Senator [John] Fertig [of Titusville] on the subject. He will be largely guided by your and Guthrie's views, and the vote will be close in the Senate." Sherman advised against repeal. He added that he would favor this only should the Pennsylvania Supreme Court be so crowded with business that it would necessitate establishment of an intermediate appellate court.[26]

In the following September he was exceptionally blunt when he urged upon state representative W. W. Wallace "the repeal or

[24] *Ibid.*, Beebe to Sherman, March 7, 17, 1873.

[25] *Ibid.*, Bates to Sherman, March 2, 11, 1874, Sherman to Shakely, February 28, [1874].

[26] *Ibid.*, Stone to Sherman, January 26, 1877, Sherman to Stone, February 1, 1877.

radical amendment of the present Bankrupt Law" of Pennsylvania. He went on to say that, having "had a personal experience with it of nearly ten years" and having canvassed the views of "the lawyers and businessmen of this part of the country, . . . in the course of my business" including "merchants of the large cities" he concluded that "there are not twenty persons of five hundred who do not consider it an unmitigated evil" aside from officials "directly benefited by the fees" accruing from "its operation." This was especially the view of "the practicing lawyers of north-western Pennsylvania."

The repeal bill had twice passed the House. Sherman added that the existing law "enables the dishonest to hinder, delay and defraud creditors, while . . . the honest debtor is subjected to harrassment, and plunder." The Supreme Court rules of practice for its enforcement were "cumbersome and impracticable." He continued: "Registers are generally incompetent and inefficient. *Ex parte* orders are the rule and not the exception, and men are condemned in their persons and estates, without hearing, and put to intolerable inconvenience and expense in endeavoring to show afterwards that they ought not to have been punished." Repeal, or at least amendment so as to secure "Registers . . . capable of filling judicial offices, and subject to prompt removal for inefficiency" and to impose "uniform rules of practice" requiring better qualified Registers "to hold stated Courts at places within reach of the people" was imperative. As for the existing law, "no creditor can afford to look after a claim of ordinary amount" under it "and few really poor debtors can command the money to pay the expenses of an uncontested bankruptcy. While a contested case is attended with unspeakable horrors of vexation and pecuniary loss." If Wallace were "familiar with public sentiment upon the subject," he would be "convinced that I have not exaggerated," [27] Sherman concluded.

At the insistence of his clients Sherman wrote to Senator Stone in March 1878 to object to a "very dangerous amendment to the limited partnership act in Senate Bill No. 207 now on its passage" at

[27] *Ibid.*, Sherman to W. W. Wallace, December 31, 1877.

Harrisburg. This was because it would delegate the management of those companies almost exclusively to boards of managers who "had no interest" in them. Senator Fertig consulted him during that session on a proposed local court bill to benefit Titusville.[28]

While active in this role of legislative adviser Sherman thus made evident his intolerance of the inefficiency and inequities of the existing legal system. This included the local courts. When Judge Lowrie died in November 1876, Sherman joined with Bloss of the *Morning Herald* in the attempt to secure the gubernatorial interim appointment of Colonel D. C. McCoy to fill the vacancy. After McCoy was appointed he presented the application of a Mrs. H. H. Rogers for the position of court stenographer of Crawford County. He was obliged to report to her the court's objection that "a lady could not get along well in reporting some of the cases . . . in Quarter Sessions, owing to the nature of the cases and character of the testimony." He added: "A criminal court is often times a rough place." As for himself he believed that "the moral effect upon the lawyers" trying such cases and upon the witnesses and spectators "of a lady's presence, industriously taking down all that was said would be good." "But still I *do* doubt whether she would not get disgusted," he informed her in partial agreement with Victorian standards. The existing court reporter of Crawford and Venango counties would hire Mrs. Rogers, he thought, if she would attend "to the *civil* business, and he to the *Uncivil* or criminal department," he added wittily.[29]

An important phase of his practice in the mid-seventies was related to the liquidation of the Titusville and other banks that had closed after the Panic of 1873. This involved advising receivers and making collections from the banks' debtors, which required an astute appraisal of the latter's character and possible courses of action. Extensions of time secured by notes backed by collateral,

[28] *Ibid.,* Sherman to Stone, March 12, 18, 1878, J. Fertig to Sherman, March 18, 1878.

[29] *Ibid.,* H. C. Bloss and R. Sherman to G. K. Anderson, November 14, 1876, Sherman to J. Trunkey, November 18, 1876, Trunkey to Sherman, November 19, 1876, Sherman to Mrs. H. H. Rogers, November 22, 1877.

such as William H. Abbott offered the defunct Citizens Bank of Titusville, depended upon whether or not he and others like him among the best debtors would go through bankruptcy before the due dates. Abbott's character, his property in Butler County, and his investment in the Grant Pipe Line Company were such, Sherman believed as the receiver's agent, that he would not resort to this method of escaping his obligations, while the prostrate condition of business in 1875 justified the extension.[30] Like Emery, Jr., Abbott eventually negotiated a settlement with the receiver.

Sherman's correspondence contains caustic references to the stockholders who were liable for the debts of the Citizens Bank. Acting for two creditors he demanded a vigorous investigation of the stockholders' assets. Some of them had been fraudulently discharged of their obligation under the bankruptcy law. Actually Sherman had, as the receiver's agent, collected all but $12,000 of "the remaining unsettled indebtedness" placed in his hands. Now he demanded action under the bankruptcy law to "prove the debt" of the stockholders and insisted that the receiver "stir up" those with whom advantageous settlements could be negotiated. Among such were Henry Harley, who could pay $10,000 he asserted.[31]

Sherman was outraged in late March 1876 when the United States District Court allowed William H. Abbott in the Citizens Bank bankruptcy case to resign as assignee when a state court's rule was pending requiring him to make an accounting and when another such rule had been issued obliging him to show cause why he should not be removed. John D. Archbold, now of the Acme Oil Company, supplied him with affidavits for use in litigation involved in collections for that bank's receiver.[32] Sherman acted in a similar capacity for the defunct Farmers and Mechanics Bank of Rochester, carrying a suit against a debtor to the Pennsylvania Supreme Court in litigation that earned him such large fees that he had a friend investigate the receiver's ability to pay. So efficient was Sherman as

[30] *Ibid.,* Sherman to Gardner, August 10, 13 [1875].

[31] *Ibid.,* Sherman to Pearson Church, March 18 [1876].

[32] *Ibid.,* Sherman to S. C. McCandless, November 28, 1876, Archbold to Sherman, February 17, 1877.

collector that he was invited to become a correspondent of The Mercantile Collection Bureau and receive its commissions.[33]

In November 1876 occurred the failure of the Peoples and Mechanics Bank in Titusville. Precipitated by the Pennsylvania Transportation Company's failure, it "created distrust of all pipe lines" and threatened a "general panic." [34]

Sherman's reports to Dun & Company that year had conveyed inside knowledge of the oil industry, as when he reported that the Acme Oil Company had leased two local refineries that had been purchased by Charles Pratt and William G. Warden, who were "connected with the Standard Oil Co." Lewis Emery, Jr., he reported, had court judgments against him "and is not good," an accurate appraisal of that producer's financial embarrassment before his operations in McKean County extricated him. Dun & Company quizzed Sherman in detail about the Pleasantville Bank, whose officers had lost heavily in the failure of the Pennsylvania Transportation Company.[35] For a year in advance of that event he had reported repeatedly on Henry Harley's increasingly shaky credit. His downfall precipitated the failure of the Producers and Manufacturers' Bank of Titusville, an event which intensified business distress in the area. Soon Sherman became involved professionally in the settlement of the complicated claims of creditors against the Pennsylvania Transportation Company.

Harley was a speculator-promoter of the Gould group. In behalf of Gould he had brought the Rockefeller group of refiners into the Erie Ring's circle and given that refining coterie its first great boost up toward industrial preeminence. However, he had been abandoned in 1871 by the Rockefellers as their mainstay in oil transportation. Then in 1872 Harley found himself without Gould's backing when that figure was ousted from the presidency of the Erie Railroad. Shortly afterward the Standard Oil entered the pipe-

[33] *Ibid.*, Sherman to Wm. H. Bowman, December 6 [1875], H. C. Albee to Sherman, December 13, 1875.

[34] *Ibid.*, Sherman to George A. Stewart, November 9, 1876.

[35] *Ibid.*, Sherman to Dun & Company, June 26 [1876], Dun & Company to Sherman, November 4, December 6, 1876, Sherman to J. Marshall, October 28 [1875].

line business by way of the American Transfer Company under Daniel O'Day's management. In 1874 its competition with the Pennsylvania Transportation Company was intensified when the Standard Oil purchased a third interest in the United Pipe Lines Company which soon became the second leading pipeline system. Because of these circumstances the pipeline pool that Harley had organized in 1874 with railroad backing was of limited value to his company.[36] The United Pipe Lines developed into a regular system the issuance of oil certificates to owners of oil in transit or storage in its lines. The certificates quickly became the basis of commercial trading in oil on the exchanges in Titusville, Oil City, Pittsburgh, and New York.[37]

Titusvillians were better informed about the vigorous expansion program of the Empire Transportation Company. Its history and program were publicized in 1876 in the *Morning Herald,* while such visitors as the Shermans at the great Centennial Exhibition at Philadelphia witnessed there its exhibition building which depicted its pipeline, railroad, and water routes of transportation that served the petroleum industry. Disruption of a short-lived pipeline pool and a bitter rate war that reduced pipage charges to five cents a barrel occurred in September, accompanied by the overissue of oil certificates by the competing companies.[38]

The Pennsylvania Transportation Company's failure was attributable not only to these difficulties but also to other factors. Rockefeller's success from 1875 to 1876 in directing the Central Refiners Association had resulted in the purchase and dismantling by the

[36] Sherman Papers, Sherman to Dun & Company, October 26, 28, 1875, November 6, 1876.

[37] In the *Titusville Daily Courier,* September 17, 1874, "Pipe Line Reports" listed sixteen pipeline companies. In the Oil Region the five companies leading in oil in storage were the United, Union, Pennsylvania Transportation Company, Antwerp, American Transportation Company, in that order, with the United far ahead. Cf. Johnson, *op. cit.,* pp. 29–30, which presents a different account of the origin of the United. But see Nevins, *Study in Power,* I, 184–185, and *Titusville Morning Herald,* June 22, 1876. For the pipeline pool, see Williamson and Daum, *op. cit.,* p. 403.

[38] *Titusville Morning Herald,* May 18, July 20, September 21, 1876.

Standard Oil of many independent refineries in the Oil Region, with a consequent loss to the Harley line of their patronage. The United Pipe Lines and American Transfer companies apparently received the patronage of the Central Refiners Association. To meet the changed situation Harley reorganized his company in July 1876, placing on its directorate representatives of six local pipeline companies. He then incorporated the Associated Pipe Line Company and began construction of a trunk pipeline from the Oil Region to salt water at Baltimore, financing it by the sale of 10% interest-bearing gold bonds and, as the marketing of these lagged, by the excess issue of oil certificates by the Pennsylvania Transportation Company. This attempt to free himself from dependence upon the trunkline railroads which had shifted their alliance from him to the Standard Oil's and Empire's pipelines had been followed by the disruption of the pipeline pool and the bitter rate war. This caught Harley's system in a vulnerable position. When the Baltimore capitalists associated with his seaboard project withdrew in fright his overissue of oil certificates (overissuing being a custom which all pipeline companies now indulged in despite the Act of 1874) their action together with the drastic decline in pipage rates and railroad nonpayment of contracted pipage to the Pennsylvania Transportation Company combined to force it into a receivership arranged by Harley.[39]

[39] *Ibid.*, April 29, July 22, 27, August 5, October 28, 30, 1876; Sherman Papers, Sherman to J. Marshall, January 8, 1876, Sherman to F. Fitz, October 30, 1876, Sherman to Dun & Company, November 6, 1876, *Paper Books*, VIII, No. 2, *Appeal of James E. Brown in the Case of George R. Yarrow, Trustee v. The Pennsylvania Transportation Company et al., Supreme Court of Pennsylvania, Western District, No. 51, October and November Term, 1880,* which shows that following the Gould precedent there had been some looting of the Pennsylvania Transportation Company by its directors, viz., Vice-President G. K. Anderson, *Paper Books,* XII, No. 5, *Argument for Defendant, Pennsylvania Transportation Company v. The Pittsburgh, Titusville and Buffalo Railway Company, Court of Common Pleas of Crawford County, Sitting in Equity, No. 79, September Term, 1880;* Johnson, *op. cit.*, pp. 259–260 n. 57; Pennsylvania Transportation Company, *Extension of Pipe Line to Tide Water* (n. p., 1876), pp. 3–12; Ralph W. and Muriel E. Hidy, *Pioneering in Big Business* (New York, 1955), p. 202.

Harley's success in arranging a receivership for his company aroused Sherman's suspicions and ire, and explain much of a later phase of Sherman's career. He was increasingly interested in the consequences of amoral business methods and was concerned at the adverse effect of the failure of the Pennsylvania Transportation Company upon the deepening business depression.

Sherman's appreciation of business integrity heightened as he worked under the unusual load of business that the difficulties of the oil men embarrassed by the depression brought to his law office. In early September 1875, for example, he was counsel in seventeen of the thirty-five civil cases scheduled for trial before the Crawford Court of Common Pleas at Meadville. Among this long list was his argument on the bill of exceptions he had filed in a collection suit against Lewis Emery, Jr. Emery's varied enterprises, which included merchandising as well as oil production, had made him the butt of successive civil suits. These arose from the struggle of individuals to survive by collecting overdue accounts from debtors whose own survival depended upon their creditors' willingness to scale down claims proportionately to the general decline in values and then to await payment until conditions improved. In another suit Sherman informed the defendants, the producing company of L. E. Emery, Jr., and E. G. Patterson, his clients, that they were "entirely wrong" in attempting to restrict payment of the oil royalty to their lessors to 12.5% instead of the contracted 33.5%. He urged that the plaintiff and defendant accept "an amicable settlement if possible, as a fight over an oil farm often results in a loss even if you win." [40]

Sherman negotiated a settlement of Emery's debt of $30,000 to the defunct People's and Mechanics Bank of Titusville so as to protect him from usurious interest and ruinous judgments if the receiver took the claim to court.[41] Much of Emery's ability to

[40] Crawford County Court of Common Pleas, *Appearance Docket,* LV; Sherman Papers, Sherman to Gault & Martin, June 26, 1873, T. C. Spencer to Sherman, March 1, 1875, Sherman to C. Heydrick, August 28 [1875], October 7 [1876], Sherman to W. Creighton Lee, August 28, 1875. During the January 1875 term of the same court Sherman had appeared in twenty-four actions, including another for Emery.

[41] Sherman Papers, Sherman to Emery, Jr., May 31, 1877, December 5, 1878.

extricate himself from the morass of debts that embarrassed him and to recoup his fortunes in the new Bradford field of McKean County stemmed from Sherman's skill in protecting his interests during successive suits and negotiations with his creditors. There developed from this invaluable service a mutual respect and personal friendship that became a leading asset for both men and one which made them an especially formidable team in future campaigns of the independent oil men in their fight for continued competition and the right to do business profitably against the rising Standard Oil.

Sherman's professional reputation brought him important litigation in behalf of outside interests. In the January 1875 term of the Crawford County Common Pleas court he won a judgment against George K. Anderson, Harley's former associate, for Edwin M. Lewis, Trustee for Jay Cooke. Sherman represented the Third National Bank of Philadelphia in litigation that continued until 1880. In behalf of a jobber he secured a writ of replevin against a local firm of jewelers. Many of these actions involved the attempts of creditors to collect overdue debts. In March he collected a debt owed to the Merchants National Bank of Indianapolis. He worked with the sheriffs in the Oil Region in the issuance of writs of attachment to oil properties in behalf of creditors for whom he had secured judgments. He prepared his witnesses carefully in advance of such civil actions, as when he acted for the Erie Iron Works against the Oil City Iron Works in collecting a debt for torpedo cases.[42]

Sherman was associated frequently with attorneys of other cities and of Titusville in his practice, whether in the trial of civil cases or in the searching of titles to oil lands, a process which tempted him to profanity, as he remarked to an attorney whom he retained to assist him. A dignified but firm courtesy toward the debtors of his clients explained much of his success in collections.[43]

He was irked that his busy professional schedule too often meant

[42] Crawford County Court of Common Pleas, *Appearance Docket,* LV; Sherman Papers, Sherman to William G. Sonn, March 1, 1875, Letterbook I, 8, 62, 65a, 68, 69, 73.

[43] Sherman Papers, Sherman to Heydrick, April 26, 1875, Sherman to R. Andrews, April 26, 1875, Sherman to Cadam & Donaghue, May 4, 1875.

hasty preparation of paperbooks (printed briefs) to support appeals to the Pennsylvania Supreme Court, resulting at times in "errors of grammar and construction." This he admitted in reply to his uncle Eleazer Conkling's praise, which he accepted modestly. During 1874 he had won an unusual number of cases before that court, many of them for the Second National Bank of Titusville in pursuit of its debtors. One important case that he won there was carried to the United States Supreme Court on writ of error in the spring of 1875. This he expected to follow to Washington the following winter. Other appeals to that high court necessitated his appearance before its bench again in 1876 and 1877. Thus, the ex-Confederate cavalryman rose to the highest level of professional practice of the American bar.[44]

Sherman exercised meticulous supervision of his law practice as revealed in his almost daily instructions to M. J. Heywang, clerk of the firm, during the summer of 1875 when Sherman vacationed with his wife on Great Lakes steamers to Duluth and by river steamboat to Davenport, Iowa. One such letter instructed Heywang: "Examine the U.S. Dist & Circuit trial list for Erie in the 3 months of July-September and see what cases I have. There is one of Wilkins & Amy which is already continued. There may be some bankruptcy case." Would Heywang also send him advertisements of sheriffs' sales of Crawford, Franklin, and Venango counties? After revealing that Sherman's training of him had enabled him to win a minor skirmish, Heywang replied, "All else is well with us—only we miss you very much."[45]

Sherman's clients trusted him implicitly, assigning judgments to him for collection, proceeds of which he paid conscientiously. The Second National Bank of Titusville gave him large discretionary power in the sale of lands foreclosed in collection actions and then bought in by him at sheriff's sales for subsequent disposition. He

[44] *Ibid.*, Sherman to E. M. Conkling, April 21, 1875, February 5, 1877, Sherman to D. W. Middleton, April 10 [1875]; *Pennsylvania State Reports*, LXXVII, 94–103, 114–118, LXXIX, 453–459, 470–473, LXXX, 163–165, LXXXIII, 203–206, LXXXV, 528–534.

[45] Sherman Papers, Sherman to Heywang, July 5, 8, 1875, Heywang to Sherman, July 12, 1875.

acted similarly for other clients, whether the collateral was land or corporation stocks, in successive attempts to minimize their losses. His maxim, which he conveyed to Heywang, "get judgment against all you can," explained his popularity with responsible businessmen and his constantly growing practice. His fees in such collection litigation for the Second National Bank were "5% on first 1,000 and 2% on balance." [46]

Such success explained why his old friend, James Sill of Erie, now President of the Erie Bar Association, invited the Shermans to attend its annual picnic on August 18, 1875, "at the head of the Bay." Only "the most urgent business" obliged Sherman to decline.[47]

Shortly after this Sherman & Beebe dissolved their law firm. Sherman continued for some years afterward to practice alone while maintaining cordial relations with his former partner. It was Beebe in January 1876 who introduced him to Justice S. B. Black of the United States Supreme Court and supported his application for admission to practice before it with the humorous observation: "The worst that I know that could be said of him is that he is a very strong *Democrat.*" [48]

Sherman's relations with his clients were distinguished by astute candor, as when he advised a Beaver, Pennsylvania, client that it was very risky to bid in lands at sheriffs' sales without knowing about the titles. He explained carefully in advance to his clients the aspects of their suits and their chances of success. He required moderate advance deposits of money to start proceedings, and after suit refunded the balance unspent together with a careful accounting. He flatly refused to take cases upon his "own account," namely, when he would be compensated *if successful* by a percentage of the damages or debt collected. He repelled the intimations of an occasional injudicious client that he was actuated in handling his case by

[46] *Ibid.,* G. C. Hyde to Sherman, November 12, 1877, Sherman to Heywang, June 30, July 19, 1875.

[47] *Ibid.,* Sherman to Sill, August 17, 1875.

[48] *Ibid.,* M. C. Beebe to J. S. Black, January 29, 1876, Sherman to Beebe, January 26, February 11, 1876.

considerations other than his interest. Sherman consistently refused to undertake litigation that conflicted with his professional obligations. He declined to institute suits when it "would do no good" in the light of existing law.[49]

So far as he could, he held public officials to an equally rigorous observation of their obligations. In May 1877 he informed the Sheriff of Crawford County that he would proceed against him if he attempted to enforce an ejection writ against the actual owners of a property. Two years earlier he had vainly instituted proceedings at Meadville to have that drunken, insolent, incompetent, and dishonest official ousted from office for malfeasance.[50]

Sherman's capacity for indignation at fraud, dishonesty, chicanery, deceit, and malfeasance extended to members of his own profession who, possessing the means to settle their debts, fought collection of them up to the state Supreme Court. From one such rogue he collected the entire amount due the Farmers and Mechanics Bank of Rochester by employing all available legal resources. Then, after this signal victory in prolonged litigation he experienced the shock of having his modest fee of $489.10 arbitrarily reduced by that bank's receiver.[51] With bored weariness he defeated the fraudulent attempts of dishonest debtors of the Oil Region and skillfully smoked them out into frightened settlements.[52] He even beat his friend and quondam client, Dr. Shamburg, President of the Chautauqua Lake Railroad.[53] As Beldon & Company's counsel he secured an injunction restraining the Shenango & Allegheny Rail Road Company, of which S. C. T. Dodd was Presi-

[49] *Ibid.,* Sherman to W. H. Gardner, April 23, 1875, Sherman to J. S. Ruton, January 4, 1876, Sherman to T. J. Rayner, February 28, 1876, Sherman to A. J. Johnson & Company, March 4, 1876, Sherman to M. C. Beebe, September 9, 1875, Sherman to Hewett & Root, March 5, 1877, Sherman to Miller & Dempsey, July 21, 1879, Sherman to A. H. Bronson, February 16, 1880.

[50] *Ibid.,* Sherman to Marsh & Harwood, August 9, 1875, Sherman to Daugherty Bross & Company, December 18, 1875, Sherman to George P. Ryan, Sheriff, May 21, 1877.

[51] *Ibid.,* Sherman to Hector McLean, December 6, 1875, December 11, 1875.

[52] *Ibid.,* Sherman to Campbell & Griffiths, August 10, 1875.

[53] *Ibid.,* Sherman to David Jones, May 8, 1876.

dent, from paying "over to any person the sum of about $255,000 due by it to the A & R W R R Co as trustee" and from issuing bonds to pay that debt, backing this up with a warning letter to Dodd.[54]

In October 1877 W. H. Kemble, President of the People's Bank of Philadelphia, retained him to assist in the appeal of a partnership case before the Pennsylvania Supreme Court. Sherman prepared his paperbook in two installments and sent it to F. Carroll Brewster, Kemble's senior counsel. It had been at Kemble's bank in September 1872 that the Pennsylvania Transportation Company's directorate had met to revise its pipage contract with the Oil Creek and Allegheny River Railroad, the nonfulfillment of which contract in 1876 contributed so heavily to the Harley combination's failure. Perhaps this later retainer, with its implied tribute to Sherman's professional reputation, was related also to his earlier legal services for that ill-fated corporation.[55]

This period of Sherman's law practice terminated in December after the great railroad riots of July had been suppressed and when the free silver and Greenback movements were gathering force to challenge the established order. Judge John Trunkey, who had admitted him to practice in 1867, was elevated in November to the Pennsylvania Supreme Court. He and Sherman had become very warm friends as the latter pled on occasion before the bench of Venango's Court of Common Pleas. Colonel S. C. T. Dodd arranged a testimonial dinner in Trunkey's honor at Franklin and invited the Shermans to attend as special guests. Unable to be there, Roger Sherman declined with a graceful tribute to Trunkey's "eminent learning and fairness," predicting that his "virtues and talents and . . . fame" would become "the common property" of the Commonwealth.[56]

[54] *Ibid.,* Sherman to Dodd, October 3, 1877, MS "Shenango & Allegheny Railroad Company."

[55] *Ibid.,* Sherman to F. C. Brewster, October 19, November 7, 1877, *Paper Books,* VIII, No. 2, *Appeal of James E. Brown,* Appendix, pp. 86–87.

[56] *Ibid.,* Scrapbook II, clipping dated Franklin, Pa., December 28, 1877, containing Sherman to S. C. T. Dodd, December 27, 1877.

# CHAPTER IV

# *The Producers and the Seaboard Pipeline*

BETWEEN 1874 and 1878 Roger Sherman won a unique professional position in Titusville and in the expanding Oil Region. While he attained leadership in the bar he also developed an intimacy with the leaders of the oil producers, a relationship the first fruit of which was state regulation of the petroleum pipelines. He did not presume upon his position, but remained aware that for fellow attorneys and others he was a newcomer to Pennsylvania and an ex-Confederate.[1] Within the Democratic party he had to remain in the background, exerting his influence with clients and professional associates in behalf of men whom he regarded as desirable nominees and of policies that he preferred, while attending the state Democratic conventions as a delegate. He was an avowed partisan, or as the Republicans of the day put it, "an unterrified Democrat."

After the Democratic victories in the October elections of Ohio and Indiana in 1876 he predicted confidently to his friend F. R. Lanier of Osceola, Arkansas, that their party would carry the presidential election in November. He warned that the Republicans intended to provoke the southern Democrats to violence and to use this as a pretext to justify bringing federal troops into play in the carpetbagger-controlled states. "The only hope of the South as well as of the Country is patient endurance for yet a little while. If the sun of November seventh sets upon a peaceful election throughout

[1] Sherman Papers, Sherman to J. Douglass *et al.,* December 9, 1878.

the South, it will rise upon a redeemed and disenthralled country," he predicted.

No one need have any fear that [Samuel J.] Tilden will not be inaugurated if elected. The rascals will not dare to go to such lengths of revolution, as to endeavor to prevent it, for they well know that such an attempt would be followed by such an uprising of the people as never was seen. There would not be in one year a republican left outside of New England. The right men would get whipped this time.[2]

Written privately as this was to Lanier, another ex-Confederate, this passage reveals that Sherman did not appreciate the determination of Republican leaders to prevent their overthrow. When he predicted to Dr. Cook of Buffalo that Tilden would carry Pennsylvania by a majority of 40,000, that family friend offered his best medical services free if Tilden beat Rutherford B. Hayes by a 5,000 majority in New York state.[3]

On November 9 Cook admitted that Tilden was elected by a close margin, claiming that he had carried the Empire State "by the most evident frauds" in the metropolis. He predicted also that Florida and Louisiana, upon which the national result depended, "will be eventually declared for Tilden tho' in both it is the shot gun that has done it." He continued: "The Democratic party must plant themselves upon the doctrine of human rights for all and the 'shotgun policy' must stop." Despite Sherman's protest Cook reiterated that it was only by "intimidation & fraud of the most glaring character" that the Democratic party had won in the South. He asserted that the Republicans intended despite this to act only "upon the absolute vote cast in the three disputed states" of South Carolina, Florida, and Louisiana, where the Republican returning boards were preparing to award the presidential election to Hayes. Sherman's view of that "stolen election," as the Democrats always regarded it, has not been preserved.[4]

He did not quit politics in disgust. Yet, if the Democrats had

[2] *Ibid.*, Sherman to Lanier, October 30, 1876.

[3] *Ibid.*, E. G. Cook to Sherman, November 4, 1876.

[4] *Ibid.*, Cook to Sherman, November 9, 14, 1876, Sherman to Lanier, December 18, 1876.

risen in armed revolt against the last-minute Electoral Commission award of the presidency to Hayes there is no doubt as to how Sherman would have acted. In 1877 he was a delegate to the state Democratic convention, where he voted to nominate Judge Trunkey for the Supreme Court. He worked successfully afterward for his election.[5]

During 1878 he worked quietly to uphold the Jacksonian antimonopolist tradition in Pennsylvania within the Democratic party. Such was his influence that the Republicans circulated false stories about him. Despite his opposition as a minority stockholder of the *Titusville Daily Courier* the *Titusville Morning Herald* succeeded in purchasing the paper from N. M. Allen during the campaign. This event reduced Sherman's sphere of influence and deprived the Democrats of a party organ in Crawford County. State Democratic aspirants now presented their rival claims to him.[6] This influential position in his party explains some of his success in securing the enactment by the legislature of bills that he drafted, *i.e.,* the mechanics-lien law and the pipeline-regulation act. These measures contributed substantially to the pragmatic development of the state's regulatory function.

Leadership within the Democratic party in Pennsylvania hardly accounted for his prominence in the great fight that the oil producers made from 1878 to 1880 against the Standard Oil and its railroad trunkline allies. It was rather Sherman's antimonopolism, his growing insistence upon socially responsible business entrepreneurship, his affiliation with Patterson and Lewis Emery, Jr., and his great professional reputation that made him the obvious choice as chief counsel of the two leading but allied groups within the producers' movement.

Sherman and his associates had gleaned new insight into the true inwardness of their era and a fresh perception of the Oil Region's

[5] Sherman, *op. cit.,* pp. 64–65.

[6] Sherman Papers, C. M. Hoover to Sherman, September 15, 1869, Wm. Mutchler, Democratic State Committee Chairman, to Sherman, September 20, 1869, Sherman to D. O. Barr, August 6, 1876, Sherman to W. R. Bole, August 6, 1878.

predicament in April 1875 when Mark Twain's play, "The Gilded Age," was staged at Titusville before "the grace, beauty, wealth and fashion" that dominated the "sympathetic and appreciative" audience.[7]

The Rockefellers' swift conquest of the oil industry gave increasing meaning to that predicament. That group was temporarily secure early in 1877 in its alliance with the trunklines and control of a majority of the pipelines, and dominant in refining as the result of the "unparalleled prosperity" of the Refiners Association and new mergers with beaten competitors. The Rockefeller coterie must have concluded that its long campaign was all but won. The shrill protests and bitter hatred of "The Octopus" that Oil Region citizens voiced was almost proof-positive of this success. Secret partnership with certain leading oil producers, who had been persuaded to enter into contractual relations by an ill-kept promise that they would benefit equally as partners with the Rockefeller group from the Standard Oil's favorable freight rates, ensured the latter firm a considerable oil supply should the independent producers again rally in opposition and impose a boycott.[8]

A further evidence of the Standard Oil's power in Pennsylvania, but substantially unknown to its competitors, was the investment in 1877 of 5.38% of its capital in Pennsylvania partnerships engaged in producing, refining, and transporting oil and 24.27% of its assets in Pennsylvania corporations.[9]

Although the Oil Region was encouraged by the independents' victory at Cleveland in the case of Scofield, Shurmer & Teagle *v.* Lake Shore & Michigan Southern Railroad Company, (a decision which secured judicial invalidation of the railroad's contractual obligation to reduce oil freight rates for the Standard Oil while

[7] *Titusville Morning Herald,* April 3, 1875.

[8] *Ibid.,* May 28, June 23, 1875, January 24, 1877; Sherman Papers, *Paper Books,* IX, No. 8, *Hascal L. Taylor, John Satterfield & John Pitcairn, Jr.* v. *John D. Rockefeller, H. M. Flagler, et al.* The "Trust" device was employed in their contract of December 1, 1874, apparently to the great profit of the Rockefeller group at the expense of the others.

[9] *Pennsylvania State Reports,* CI, 125, Commonwealth *v.* Standard Oil Company.

charging higher rates to its competitors), it was obvious that the independent producers and refiners of northwest Pennsylvania had to devise their own remedy if they were to restore business freedom and secure fair prices for their products.[10]

In an attempt to accomplish this, the independent producers, led by B. B. Campbell, a large producer of Parnassus, persuaded the Pennsylvania Railroad to terminate its alliance with the Rockefellers in the early spring of 1877 by pledging to ship on that carrier as large a volume of crude oil as the Standard Oil had shipped. The Pennsylvania's bolt from the trunkline oil pool and discontinuance of its favoritism to that business giant precipitated a bitter-end rate war. In this the northern trunklines, backed by the United Pipe Lines Company, cut rates drastically. The terrible antirailroad riots in Pittsburgh in July during the great railroad strike dealt the Pennsylvania Railroad a severe blow after months of heavy losses and contributed to its defeat in the rate war. In a complicated deal on October 17, 1877, it sold the Empire Transportation Company's pipelines and refineries to the Standard Oil for $3,400,000 and its oil tank cars to a car trust in which that corporation was a heavy stockholder. This was accompanied by a secret agreement that restored the Pennsylvania to the oil freight pool and obligated it to pay large rebates to the American Transfer Company and to concede to the northern trunkline allies of the Standard Oil 58% of the oil traffic. All this was staggering proof of the power and financial resources of the restored monopoly.[11]

The Oil Region's leaders and the petroleum producers were now confronted with the might of an entirely new type of monopoly. This combined a preponderance of manufacturing capacity in the refining industry with control of the pipeline system of the Oil Region and support of the three trunklines serving it. Their discriminatory freight rates favoring the Rockefeller combination enabled it to launch an immediate drive to oblige the remaining independent refineries of western Pennsylvania to sell out to it on its own

[10] *Ohio State Reports,* XLIII, 574–613.

[11] Nevins, *Study in Power,* I, 231–249. The allegedly extenuating circumstances, pp. 266–267, are open to grave exceptions.

terms. Victory in this would make the producers completely dependent upon the market provided by the Standard Oil. Actually, with its subsidiaries, that company was almost the sole purchaser of crude oil while the pipelines of the United Pipe Lines Company provided the only means of transportation from a large majority of the wells.

The independent oil men of the Oil Region resorted to other tactics to restore the competitive market. One possibility was to construct independent pipelines, either to independent railroads or to the Great Lakes-Erie Canal route, or to build the seaboard pipeline that the producers had demanded in 1875 and Henry Harley had promised in 1876. Another tactic was to invoke the power of the Commonwealth of Pennsylvania in behalf of restoration of free enterprise in the Oil Region by enforcement of the common-law obligations of the railroads and pipelines. This could be done by new legislation and by litigation instituted by the Attorney General.

Among the producers two groups appeared. Following the leadership of Lewis Emery, Jr., the Equitable Petroleum Company, Ltd., began a pipeline from Bradford to Buffalo. Its initial section developed a summer route to salt water by making deliveries to an independent railroad that carried the oil to Lake Erie, whence it was conveyed via the Erie Canal and Hudson River to New York harbor. This was countered immediately by competitive rate-cutting on the northern trunklines and by their refusal to provide a winter outlet for the oil when the Erie Canal closed.[12]

More official was the project approved on December 21, 1877, by the General Council of the newly revived Petroleum Producers Unions, six weeks after the Pennsylvania Railroad's surrender to the Standard Oil-northern trunkline combination. This was the plan to construct for the producers a seaboard pipeline to Baltimore, a project recommended by the Committee on Transportation to the General Council's meeting in Titusville at the urging of B. D.

[12] *Ibid.*, I, 250–267, 289, 300–301; *Titusville Morning Herald,* August 10, October 18, 1878; Tarbell, *op. cit.*, I, 214, 222–223; Sherman Papers, Sherman to Equitable Petroleum Company, Ltd., July 1, 1878.

Benson and David McKelvy, who had until recently managed the Columbia Conduit Company, the first trunk pipeline from the oil fields to a major refining center—Pittsburgh. Having acquired a right of way for the new project they agreed to sell this to the Seaboard Pipe Line Company, Ltd. (capital $1,000,000) which the General Council voted to organize to construct and operate its pipeline. The new pipeline was to provide "equal" facilities to all members of the Producers Unions and to charge not more than $1 a barrel for transporting crude oil to salt water. This project would serve to defeat the Standard Oil-trunkline alliance and the existing blockade on non-Standard Oil shipments at a time when the rapidly increasing Bradford field production was threatening to overflow available storage tankage.[13] When the General Council received the approval of its members in January 1878 it appointed its President, B. B. Campbell, one of the two Directors to represent the producers in the new company.[14]

Benson and McKelvy promptly secured Roger Sherman as their general counsel. As compensation for the work that they asked him to do, which included clearing the titles of the right of way, he stipulated a relatively modest salary of $5,000 plus travel expenses. He joined in planning connecting pipelines in the Oil Region. He advised the managers that the right of way to Baltimore be located "with absolute definiteness as soon as practicable, and also that the proper deeds and conveyances should be made, and the abstracts of titles prepared, and perfected."[15] He was an expert upon such matters. Simultaneously, he advised Dr. Shamburg regarding his proposal to lay a pipeline along the Buffalo & Southwestern Railroad Company's track to Buffalo. Shamburg secured his contract permitting this, but Rockefeller's personal intervention with the Erie Railroad killed the project.

For the Seaboard line Sherman had to help to resolve such

[13] Sherman Papers, circular letter to local Unions from the General Council, December 21, 1877; Williamson and Daum, *op. cit.*, p. 438.

[14] *Derrick's Handbook*, p. 294.

[15] Sherman Papers, Sherman to Benson and McKelvy, January 28, 1878, Sherman to Benson, February 4, 1878, Sherman to R. E. Hopkins, February 5, 1878.

difficult problems as how to secure passage over the Juniata River at its junction with the old Pennsylvania Canal, whose bed had been purchased by the Pennsylvania Railroad Company. To General H. Haupt, chief engineer who had won fame in the Army of the Potomac, he gave specific instructions on how to locate the right of way exactly. Sherman advised on how to secure the right of crossing the tidal Patapoco River. He informed R. E. Hopkins, the third manager of the Seaboard, that under the Wallace Corporation Act of 1874 a company could be organized to construct and operate a telegraph line, such as was essential to the operation of the Seaboard line, "anywhere in Pennsylvania, and right of way taken by paying damages." [16] As general counsel of the Seaboard he required of attorneys retained in each county through which the right of way passed that they send all abstracts of title essential to it to him or to the company's main office at Titusville. He directed the necessary local litigation while informing them of their fees.[17] Recognizing the Seaboard's financial difficulties, he suggested that it credit his retainer toward his annual salary.[18]

To facilitate construction of the Seaboard line and the Equitable line he drafted a free-pipeline bill for presentation in the legislatures of Maryland, New York, and Pennsylvania. He revised meticulously the Pennsylvania Senate bill No. 245 that was based upon his. The producers, he advised Senator Stone, "are very desirous that we shall pass it." However, it was defeated in the House of Representatives on May 8. Simultaneously he stood guard against proposed changes in the Pennsylvania law authorizing limited partnerships under which "some of our friends expect to do business." Thus he protected the Seaboard's pending application under the limited partnership act of 1874.[19]

[16] *Ibid.,* Sherman to Haupt, February 25, March 25, 1878, Sherman to A. P. Stevens, February 25, 1878, Haupt to Sherman, January 28, 1878, Sherman to R. E. Hopkins, February 11, 1878.

[17] *Ibid.,* Sherman to ?, March 25, 1878, Sherman to F. P. Stevens, April 22, 1878, Sherman to S. Burke, August 12, 1878.

[18] *Ibid.,* memorandum, Sherman to Seaboard Pipe Company, Ltd., January 26, 1878.

[19] *Ibid.,* Sherman to S. W. Stone, March 9, 12, 1878, Sherman to John Fertig, March 12, 1878, Sherman to F. P. Stevens, April 4, 1878.

At Annapolis the Seaboard's Baltimore attorney, F. P. Stevens, secured permission from the Maryland legislature for the crossing of the Patapoco. With the support of the Baltimore and Ohio Railroad he secured certain other legislation despite the "stealing" of the free-pipeline bill. But when Stevens submitted his bill for $10,800 for expenses Sherman protested bluntly to his Board of Managers and then informed Stevens of its refusal to pay "the bill or any part of it." The company's policy, he explained, was not to pay or promise any money for legislative purposes since doing so would make it vulnerable to future legislators' blackmail. It was "opposed upon principle to the use of money to effect legislation," and it had not authorized the expenditures or suspected that "any such liability was being incurred." [20]

Made toward the close of the most corrupt decade of "The Gilded Age," the Seaboard's declaration to its Baltimore attorney was a significant innovation in entrepreneurial method. As a contemporary might have perceived, it expressed the Tilden reform influence in a manner that contrasted vividly with the methods employed by the Rockefellers, Thomas A. Scott, Gould, and the Vanderbilts. The oil producers' revolt from industrial and transportation monopoly was, then, an ethical as well as a business rebellion from the amoral practices of the era.

Benson's connection with the Baltimore and Ohio Railroad, which had shipped the Columbia Conduit's oil to Baltimore from Pittsburgh before that pipeline had been sold to the Standard Oil by Dr. David Hostetter in October 1877, explained the Seaboard's legislative success at Annapolis. That explained also the Benson-Hopkins-McKelvy trio's willingness to team with the Producers Unions in constructing the Seaboard line to Baltimore, in a joint attempt to reopen competition in oil buying, shipping, and marketing.[21]

Although Sherman worked to secure clear titles and valid leases for the right of way, informed General Haupt of defects in legal papers with a view to their perfection, spotted gaps in the right of

[20] *Ibid.,* Sherman to F. P. Stevens, May 2, 1878.

[21] Johnson, *op. cit.,* pp. 50, 65–66; Nevins, *Study in Power,* I, 203–248; Tarbell, *op. cit.,* I, 194–195, 198.

way accurately and instructed the local attorneys on how to perfect the company's titles, the line to Baltimore was not laid. Despite difficulties, rights of way were secured in the four western counties so that the Seaboard was able to lay its gathering and connecting lines in the Oil Region in a manner that ensured it an adequate volume of oil traffic, but strategic locations were unobtainable on the route to Baltimore, and funds were insufficient to complete the pipeline to that port. In late September 1878 Stevens, the Seaboard's attorney there, asked Sherman to inform him of "the prospects of the Pipe Line." [22] Sherman had secured the Seaboard's charter in May, paid for rights of way deeds and leases, and protected lessors when the Pennsylvania Railroad attempted to eject them from lands adjoining the Old Portage Railroad whose right of way the Seaboard intended to use. The managers, though, were "embarrassed by many unlooked for things," he informed Stevens; yet, at the same time, he advised the company's inquiring local attorneys that he had "no information that the Seaboard Pipe Line has been abandoned. On the contrary I hear that all things are favorable." [23]

However, a sudden loss of confidence in the project swept the Oil Region. All but eight of the original subscribers to the Seaboard's stock withdrew their money. Of those eight, five were "friendly, one doubtful & two not within reach," Secretary-Treasurer Hopkins informed Sherman on October 24. The managers were unable to pay Sherman his full salary that month, but he continued on as General Counsel. At a special Trustees' meeting those who had subscribed $5,000 each for stock were asked to transfer it back to the company. The Seaboard pipeline project faced catastrophe.[24]

[22] Sherman Papers, F. P. Stevens to Sherman, September 25, 1878, Sherman to Haupt, April 24, May 3, 1878, Sherman to J. G. Lake, April 24, 1878, Sherman to Krebs & Hindman, April 24, 1878, Sherman to J. A. C. Ruffner, April 24, May 2, 1878, Ruffner to Sherman, May 17, 1878.

[23] *Ibid.*, S. B. Stewart to Sherman, May 20, 1878, J. G. Lake to Sherman, July 31, 1878, Sherman to F. P. Stevens, September 30, 1878, Sherman to Ruffner, September 30, 1878.

[24] *Ibid.*, Hopkins to Sherman, October 24, 1878, Sherman's bill to the Seaboard Pipe Line Company, Ltd., December 31, 1878.

If Sherman had needed intimate insight into the difficulties experienced by the producers he acquired it from producing oil on a limited scale in Millerstown in McKean County in partnership with his mother-in-law, Mrs. C. B. Seymour. This was the result of having extended financial assistance to the Seymours after the Panic of 1873, when he kept his father-in-law on as manager after purchasing his share in the family partnership. During 1878 Sherman corresponded with him regularly, studied the oil market, and instructed him on how to deal with the "immediate shipment" policy followed by the United Pipe Lines Company. This policy, instituted arbitrarily during the glut in Bradford production in a manner that subjected independent producers and oil buyers and shippers to an especially onerous squeeze, was made possible by United Pipe Lines' control of railroad tank cars. Rockefeller resentment at the Oil Region's prolonged resistence to his monopolistic policy and at the billingsgate showered upon Standard Oil executives now found expression in a vindictive vengeance upon the independents.[25]

Sherman needed the income from his oil. He resented the United Pipe Lines' practice of buying and running his oil and then paying for it only after months of delay, a policy that obliged the producers in effect to supply operating capital to that monopoly. Sherman's insistence that his wells' operating expenses be such as to net him more than $500 a month indicates how limited were the operations of Sherman and Seymour.

The abuses of "immediate shipment" buying by the United Pipe Lines aroused Sherman's intense resentment. Among these was the 20% reduction in the purchase price from the market price for oil and the delays in running oil from his storage tanks. Seymour thought it better to "let the oil be in the ground than get it out" and sell at $1 a barrel. He had to submit, however, but was unable to persuade United Pipe Lines to run more than 100 barrels of their oil a day. Kendall Creek, where their wells were located, was not in the flush producing district. There the operators were obliged to run their oil on the ground and to let it go to waste for weeks when storage tanks were full, pending United Pipe Lines' decision to

[25] *Ibid.*, Sherman to C. B. Seymour, January 16, July 1, September 30, 1878.

purchase their oil at "immediate shipment" prices and then to run the oil at its convenience. Overproduction, intensified by reckless drilling in the McKean County field, enabled the Standard Oil to capitalize fully upon the advantages of its control of pipeline transportation.[26]

The construction by United Pipe Lines of additional oil tankage, a service for which it now charged triple rates despite the decline in oil prices, lagged behind the demand for storage. The 20% discount of "immediate shipment" crude-oil prices brought huge profits to the Standard Oil. It was this situation that had persuaded the General Council of the Producers Unions, prodded by Emery and Benson, to continue to support the independent pipeline projects as the means of securing a competitive market price for oil and of terminating the abuses of the United Pipe Lines' management.[27]

Sherman's loyalty to the Seaboard pipeline had been intensified in July 1878 when the Standard Oil bought the Buffalo & McKean Railroad Company and cut the Equitable pipeline off from rail transportation to Lake Erie, and again when United Pipe Lines renewed its "squeeze" in oil prices in January 1879. To "get even in the oil operations" he resorted to operating four wells with two sets of machinery. Like other producers he began to drill in a new producing district, thus seeking a remedy for the evils of overproduction by increasing his firm's production. He employed the customary torpedoes to keep wells producing at a maximum and ordered Seymour to sell the oil "every thirty days if possible," remitting receipts above expenses.[28]

The Seaboard was behind in paying his salary. Generously he

[26] *Ibid.*, Seymour to Sherman, July 2, 3, 9, 1878, Sherman to Seymour, July 1, August 3, 1878; Johnson, *op. cit.*, pp. 70–71; *Titusville Morning Herald,* May 15, 16, 1879; Nevins, *Study in Power,* I, 290–292; Williamson and Daum, *op. cit.*, pp. 383–390, for the best account of "immediate shipment."

[27] Johnson, *op. cit.*, p. 71; Nevins, *Study in Power,* I, 292; Tarbell, *op. cit.*, I, 216–218; Williamson and Daum, *op. cit.*, pp. 383–385. It was noted that Daniel O'Day had inaugurated "immediate shipment" after the General Council of the Petroleum Producers Unions had voted to support the Seaboard pipeline. Nevins, *Study in Power,* I, 294, 296.

[28] Sherman Papers, Sherman to Seymour, November 15, 1878, January 27, 1879.

advised Hopkins that this account could be settled in a manner satisfactory to the company. He stressed to Hopkins the importance of organizing a producer-owned storage company for the accommodation of the "large class of small producers . . . whose rights and good will should not be disregarded." If controlled by the producers themselves it would be the "first step" toward adequate protection of their interests. Obviously he intended such a firm to operate in conjunction with the Seaboard pipeline.[29]

From its inception the Benson group and Sherman may have doubted the ability of the General Council and its supporters to finance construction. The producers' desertion late in 1878 disclosed the weakness of that individualistic, rather volatile, and hard-pressed element of the Oil Region's business community. During March, months before this happened, three Titusville men had approached Franklin B. Gowen, president of the Philadelphia & Reading Railroad Company, with a proposal that the pipeline be constructed to connect with its tracks at Williamsport on the upper Susquehanna River. Secret negotiations were actually facilitated by the withdrawal of producers' support from the route to Baltimore.

Gowen agreed that the Reading should subscribe $250,000 toward the $625,000 capital of a new company. Entitled the Tide Water Pipe Company, Ltd., it purchased the Seaboard's incomplete system in the Oil Region for $125,000 and also the Equitable pipeline. With Sherman as general counsel General Haupt quickly acquired a right of way from Bradford. The main pipeline to Williamsport was completed on May 22, 1879. At last, with the aid of the "Ruler of the Reading," who also headed the anthracite-coal railroad pool, the independent oil men of the Oil Region gained an outlet to the seaboard that partially freed them from the monopolistic control of the Standard Oil and its railroad allies.[30]

[29] *Ibid.*, Sherman to Hopkins, January 28, March 3, 1879.

[30] Francis M. Buente, *Autobiography of an Oil Company* (New York, 1923), pp. 17–19, 21–23; *Derrick's Handbook*, p. 300, 306–308; *Titusville Morning Herald*, March 31, 1879; Johnson, *op. cit.*, pp. 74–75; Nevins, *Study in Power*, I, 346; Marvin W. Schlegel, *Ruler of the Reading: The Life of Franklin B. Gowen* (Harrisburg, 1947), pp. 182–184; Williamson and Daum, *op. cit.*, pp. 440–443.

Benson's spectacular achievement attracted widespread interest. The Tide Water offered the United Pipe Lines the most powerful opposition that had developed since the Standard Oil purchased the Empire Transportation Company's facilities and the Columbia Conduit Company pipeline. Bitter court battles with the Pennsylvania and other railroads under whose tracks the Tide Water's pipes were laid in creek beds, defeat of the Standard Oil's blocking attempt by discovery of a usable gap in its north-south right of way, and the development of powerful pumping stations able to pump oil over the Allegheny divide were legal, engineering, and construction feats of considerable magnitude. Allan Nevins ranks the completion of this first long-distance pipeline as an engineering feat equal in importance to the construction of the Hoosac Tunnel.[31]

Although Haupt is properly credited with the engineering and construction triumph, it was Sherman who established the legality of his coups and fought the company's legal battles successfully against the Pennsylvania Railroad in the courts. Undoubtedly it had been Sherman who advised Benson to acquire secretly the right of way to Williamsport while ostensibly at work on the route to Baltimore, but not to register the Tide Water's deeds and leases until the "opposition" had purchased farms in its attempt to defeat the project. The Rockefellers had been frightened into obstructionism by Haupt's estimate that oil could be pumped to tidewater over the new pipeline at a cost of five cents a barrel. Sherman also represented the Tide Water at Harrisburg during the legislative session.[32]

Shortly after the organization of the Tide Water Pipe Company, Ltd., on November 13, 1878, Sherman had subscribed to $3,000 of its stock. President Campbell of the General Council of the Producers Unions also became a stockholder. Sherman must have found his

[31] *Titusville Morning Herald,* April 9, May 29, June 3, 1879; Nevins, *Study in Power,* I, 346, 435 n. 3.

[32] Sherman Papers, Sherman to Benson, April 25, 1879; Johnson, *op. cit.,* p. 75; Nevins, *Study in Power,* I, 347–348. Sherman preserved the code telegrams that kept the Tide Water's legal activities, construction, and early operations secret from the Standard Oil.

professional association with President Benson especially agreeable, since the laying of the pipeline to Williamsport was achieved without resort to a construction company and was completed at a cost of less than $800,000 including the pipelines and rights of way of the Equitable and the Seaboard. This was in itself evidence of the integrity and efficiency of the management.[33]

On May 5, 1879, before completion of the line to Williamsport, Sherman informed Dr. E. G. Cook of Buffalo that the Tide Water offered promising investment opportunities. Small investors in it were not in danger of being sold out, and Benson and McKelvy, he told Cook, would "protect your interests in any event."[34] This was informed tribute to their integrity, based upon association with them in the Seaboard and Tide Water projects.

Undoubtedly Sherman thrilled with all Titusville and the independent oil men when the two large Holly steam engines pumped oil in the Tide Water pipeline to the crest of the 1,900 foot Allegheny divide whence it flowed by force of gravity to Williamsport, to be carried by the Reading's new tank-car fleet to the coast. The Tide Water's success was not only a technical triumph. It was an outstanding illustration of entrepreneurial innovation fostered by business competition, and a victory as well for the higher business ethics that the independent producers and pipeline men advocated.[35] The Benson-Hopkins-McKelvy group's competitive effectiveness was demonstrated when, as they completed the pipeline, they learned that six of their seven refinery customers in the New York area had been purchased by the Standard Oil. With the help of S. O. Brown, a banker and investor in the company, they erected storage tanks and then constructed refineries on Delaware Bay at South Chester, Pennsylvania, and at Bayonne, New Jersey, near the plant of Lombard & Ayres. Subsequently, to prevent the United Pipe Lines from cutting off its supply of crude oil, the Tide Water

[33] Sherman Papers, S. O. Brown to Sherman, August 8, 1882, B. B. Campbell to Sherman, July 2, 1882; *Titusville Morning Herald,* May 4, 1879; Johnson, *op. cit.,* p. 75.

[34] Sherman Papers, Sherman to Dr. E. G. Cook, May 5, 1879.

[35] Buente, *op. cit.,* pp. 21–23; Nevins, *Study in Power,* I, 347.

purchased producing properties in the Oil Region. After burying its pipeline underground in 1880 and perfecting methods of cleaning it, the Tide Water found the operating costs of its six-inch pipe and triplex pumps were lower than that of any of their competitor's for many years.[36]

The petroleum producers hailed the Tide Water's completion which ensured them "cheaper rates of freight" and curtailment of "the profits of the Standard monopoly in rebates and drawbacks from the railroad companies," and at the same time boycotted the *Oil City Derrick* which had come under Standard Oil control and was now abusing them and the Benson group.[37] In October 1879 Sherman and Seymour, oil producers, terminated their contract with the United Pipe Lines. They then leased their storage tanks to the Tide Water, connected their wells to its pipes, and contracted to deliver their oil to it.[38]

The anticipated higher price for crude oil was not realized in the autumn of 1879, because of the rising Bradford production, despite the result of the immediate bitter rate war which the New York Central, Erie, and Pennsylvania railroads and the United Pipe Lines waged against the Tide Water-Reading alliance, albeit against Rockefeller's better judgment. The Philadelphia & Reading Railroad Company and the leading producers associated with the Tide Water, who had supported it primarily as a means "of establishing equal rates and fair trade," accepted a moderate loss and fought back effectively, while their trunkline opponents alone suffered a loss of $10,000,000 during the rate war.[39] After Benson rejected the Standard Oil's offer to purchase the Tide Water, employ him, and purchase all oil carried by the Reading, the Rockefeller organization

[36] Buente, *op. cit.*, pp. 20, 25–26, 29.

[37] *Titusville Morning Herald,* June 9, 1879.

[38] Sherman Papers, Sherman to H. M. Hughes, October 13, 1879, Sherman to C. B. Seymour, October 13, 24, 1879.

[39] Buente, *op. cit.*, p. 23; *Derrick's Handbook,* pp. 314, 318–320; *Argument of Mr. Franklin B. Gowen in the Matter of the Investigation of the Standard Oil Trust by the Committee on Manufactures of the House of Representatives* (Philadelphia, 1889), pp. 12–15; *Pittsburgh Daily Dispatch,* January 9, 1879, quoted by Johnson, *op. cit.*, pp. 75–76.

began construction secretly of a far-reaching system of long-distance crude-oil trunk pipelines of its own to Cleveland and New York City. At the same time Standard Oil-affiliated producers attacked Emery and withheld support from the program of controlled crude-oil marketing followed by the Producers Unions.[40]

Contrary to Allan Nevins' findings it was the Standard Oil in the person of John D. Archbold that sought peace from the Tide Water as early as October, 1879, after railroad crude-oil freight rates from the Oil Region to the seaboard had declined to 10¢ a barrel. This overture was made by Archbold to Sherman as part of an attempt to negotiate a general settlement of differences between the Standard Oil-trunkline-railroad combination and the independent oil men, a settlement in which the Commonwealth Suits that Sherman directed were to be included. Sherman sent Archbold to Benson, for a meeting which produced an early and "thorough understanding of our mutual positions," Archbold, the President of Acme Oil Company, informed Sherman afterward. As for a settlement of differences between the Tide Water and the Standard Oil-trunkline combination, Sherman advised that Archbold "proceed as speedily as possible to arrange whatever differences of opinion there may be as to matters of detail in the matter suggested at the interview, and with which I have no power to meddle."[41]

This kind of advice explained Sherman's reputation for astute professional guidance of his business clients. Apparently the Standard Oil had been injured more in pride or in purse than Nevins has supposed, or it was intervening to extricate its trunkline allies from an impossible predicament. Because of the Tide Water's losses in revenue, however, Benson was obliged to negotiate with Rockefeller formally in January 1880. This produced in the spring an informal oil-traffic pool that allocated 16.66% of the traffic to the Tide Water-Reading. In addition to this evidence of Benson's and Gowen's competitive effectiveness, the agreement established new oil

[40] Johnson, *op. cit.,* p. 76; Nevins, *Study in Power,* I, 348–355.

[41] Nevins, *Study in Power,* I, pp. 350–355; Sherman Papers, Archbold to Sherman, October 10, 1879, Sherman to Archbold, October 13, 1879.

tariffs, at 51¢ a barrel to New York (instead of 80¢ of the preceding spring) and 41¢ to Philadelphia and Baltimore from common points in the Oil Region, and secured to the Tide Water large profits and to the railroads but a slender margin above costs.[42]

Thus, Benson, Gowen, and the oil producers and investors associated with them won a signal victory. Not until 1883 did the Tide Water enter into a formal oil-traffic pool with the Standard Oil and trunkline railroads, when it was granted 18.5% of the business, the portion that it had enjoyed in 1882. This resulted from acceptance of a loan from the Standard Oil at a time when the Tide Water was hard pressed after the Solar Oil Company had capitulated to Rockefeller. The pool technically restored monopolistic control of pipeline and railroad oil freight rates and terminated officially the Tide Water's competition for a larger share of the oil traffic. However, it had first breached the almost complete monopoly of oil buying, shipping, and refining that the Rockefeller organization had achieved in November 1877. The producers benefited from the lower pipage and railroad freight rates on oil that resulted from the renewal of competition, which in turn benefited the entire economy of the Oil Region as the long business depression ended. Furthermore, Oil Region capitalists in alliance with a minor railroad had obliged the Standard Oil to concede officially their right to engage in the oil business independently of the powerful combination.

Benson did not foresee, however, that after Gowen's final deposition from the presidency of the Reading that railroad would sell to the Standard Oil its large minority holding in the Tide Water. This obliged its managers, despite their evident dislike of the Rockefeller organization, to pursue friendly relations with it thereafter. Gowen, in consequence, severed connections with the Tide Water and contributed personally to the mounting literature of criticism of the Standard Oil. Certainly, after 1883, the Tide Water was neutralized

[42] *Derrick's Handbook,* pp. 314, 321; Johnson, *op. cit.,* p. 121; Nevins, *Study in Power,* I, 355–380; Schlegel, *op. cit.,* pp. 242–243; Williamson and Daum, *op. cit.,* pp. 445–446.

as a competitor and the free pipeline act that it secured that year was an empty victory.[43]

Roger Sherman's contribution as general counsel to the Tide Water's achievements before it was obliged to pool with the Standard Oil gave him a genuine share in its victories for private enterprise and competition. This greatly enhanced his prestige and influence in the Oil Region.

[43] Henry Demarest Lloyd Papers (State Historical Society of Wisconsin), Franklin B. Gowen to Lloyd, June 10, August 20, 1889; *Derrick's Handbook,* pp. 358–359; Johnson, *op. cit.,* p. 121; Nevins, *Study in Power,* I, 375–380.

# CHAPTER V

# *Launching the Commonwealth Suits*

SHERMAN continued his practice while serving as general counsel for the Seaboard and the Tide Water pipelines. Some of his work dealt with problems of the petroleum industry and tightened his bonds with the independent oil men who consulted him, for instance, regarding the organization of their companies under the limited partnership act. On February 24, 1879, he advised Emery's Equitable Petroleum Company, Ltd., in which 200 producers held stock, that its organization was "very unsafe" for its members and recommended thorough reorganization under the act. Among the advantages of doing so, he intimated, would be an ability to prevent transfers of stock whereby otherwise control of the enterprise might be lost by the organizers and the company be cynically wrecked by outsiders. His advice against incorporation was an undoubted factor in that firm's decision to merge with the Tide Water in April.[1]

More important to Sherman's education into the realities of the petroleum industry and the dominant system of business and political power in Pennsylvania than his work as general counsel of the Seaboard and Tide Water pipeline companies was his parallel service as chief associate state counsel in the famous Commonwealth Suits. This service developed from his position as chief counsel of the Petroleum Producers Unions, a position he had held since December 1877. Prosecution of the Commonwealth Suits was the third method whereby the independent oil men sought to restore free enterprise in the petroleum industry.

[1] Sherman Papers, Sherman to S. P. Boyer, February 24, 1879, "P" to Sherman, March 24, 1879.

The decision to initiate the Commonwealth Suits was a result of the activity of the General Council of the Petroleum Producers Unions, where Sherman, B. B. Campbell, the President, and Elisha G. Patterson formed a powerful triumvirate that, with the backing of Lewis Emery, Jr., and Benson's group, dominated the producers' movement. The General Council was organized in a three-day meeting at Titusville between November 21 and November 23, 1877, three weeks after the *Oil City Derrick* announced that the Standard Oil's newly achieved monopoly of the pipelines would place the producers at its mercy because of its control of the market and of oil in storage.[2] After Campbell was elected President, the General Council authorized the appointment of certain committees. Members of the committees on Legal Remedies and on Transportation performed much important work during the next two and a half years; Emery, Jr., was an important figure on the first committee and Patterson equally so on the second. As chief counsel Sherman worked closely with Campbell and counseled both committees. His warm personal ties with Emery and Patterson were as important as his official position in enabling him to influence policy development. Henry Byrom of Titusville, Secretary of the General Council, was also a friend and client of Sherman.[3]

The prominence of Campbell and Patterson in the General Council derived from their leadership in the spring of 1877 when they had persuaded Thomas A. Scott to break the Pennsylvania Railroad's alliance with the Standard Oil, withdraw from the trunkline oil-traffic pool, and ally with the independents. They had negotiated the latter's abortive agreement of July 1877 with Scott. They had led also in the protest against the Pennsylvania Railroad's surrender to the Standard Oil and northern trunklines in October

[2] *Derrick's Handbook,* pp. 288–290.

[3] Henry Demarest Lloyd, *Wealth Against Commonwealth* (New York, 1894), pp. 169–170; *Report of Proceedings before the Committee Appointed by the Pennsylvania Legislature to Inquire into the Legal Relations of the Standard Oil Company of the State* (Harrisburg, 1883), pp. 432, 441; *Appeal to the Executive,* pp. 22–24.

and November. It was they who had bestirred the Producers Unions into revived activity that had led to the organization of the General Council.[4]

On November 22, 1877, the *Titusville Morning Herald* advised the Council to reduce production or retire the stored surplus oil from the market. The General Council promptly added this objective to its policies in the hope of being able to restore crude-oil prices. Its activity paralleled that of the New York Chamber of Commerce, whose campaign against railroad discriminations and demand for railroad regulation were stimulated partially by the program of the Producers Unions. The latter organization was the first in a primary industry to demand government regulation so as to safeguard competition and the general welfare.

The best introduction to the General Council's work and achievements during these years is provided by the rare *History of the Organization, Purposes and Transactions of the General Council of the Petroleum Producers Unions* which Sherman wrote anonymously for that organization in 1880.[5] This, however, must be supplemented by the rich resources provided by Sherman's personal papers and other sources before the extent of his contribution to the partially realized program of that movement can be appreciated.

The General Council was set up as a protective organization for the producers, to gather statistics on the petroleum industry and to protect them "against all conspiracies formed to injure them in their business." The "particular" injury that the Producers Unions were organized to prevent was "the overweening control of the oil business by the Standard Oil Company." The General Council comprised 160 to 170 delegates from the local unions of the independent producers that existed in all districts of the Oil Region. It met monthly at Titusville, the organization's headquarters. Any producer could join a local union, but only the most courageous and

[4] Sherman Papers, B. B. Campbell and Elisha G. Patterson to Thomas A. Scott, President of the Pennsylvania Railroad Company, September 11, 1877.

[5] Titusville.

financially strong did so. The total membership ranged from 2,500 to 4,000 and represented invested business capital totaling hundreds of millions of dollars, a sum surpassing the total resources of the Rockefeller organization.

The great struggle that ensued, of which the Seaboard pipeline project was a part, was a contest between the competitive and monopolistic wings of the oil industry, in which the greater numbers and wealth of the former were pitted against the rigorous discipline, unity, and lawless determination of the latter. As the struggle developed, large producers with "Standard interests and affiliations" became "deadly opposed" to the General Council and the Producers Unions. Other producers who feared "the power of the Standard" maintained that yielding to it was the only means to "redress." Campbell was undoubtedly correct when he maintained, however, that "the large majority of the producers of Western Pennsylvania" were "heart and soul" behind the General Council's program. Antimonopolism was now endemic in northwestern Pennsylvania.

For eighteen months the members of the Producers Unions were oath-bound to secrecy regarding their proceedings as a means of protecting themselves from business retaliation. Oil refiners and oil brokers were excluded, except those who were engaged in oil production like Emery, Jr. who was active in refining, pipeline operation, and production. Businessmen engaged in other fields, such as C. E. Hyde, President of the Second National Bank of Titusville, were also admitted if they were engaged in oil production. Sherman's partnership with Mrs. Seymour had gained him early admittance.

The General Council's monthly meetings were not public, although its formal resolutions and many other actions were reported to the press. Deliberations were secret because the delegates did not want reporters present at business meetings. All business of the Producers Unions originated with the General Council and was ratified or not as members saw fit in formal referenda, although Campbell's lead was generally followed. Some early policies approved of in this manner were the support of Equitable's pipeline-rail-Lake-canal route and the organization of the Seaboard Pipe Line Company, Ltd.[6]

The Committee on Transportation secured the approval of the General Council and of the Producers Unions for Patterson's original "antidiscrimination" railroad or "Interstate Commerce" bill. Sherman's recommendation that it be approved had been preceded by careful study of Massachusetts' railroad statutes and reports of its Board of Railroad Commissioners.[7] He apparently helped to draft a state antidiscrimination bill that was presented to the legislature at Harrisburg. It passed the Senate there before meeting "sudden death in the lower House." There it was said that "few were able to understand it," revealing a typical "want of intellectual capacity" and "natural obtuseness" which Sherman identified acidly in his *History.*[8]

Sherman drafted a stringent bill providing for regulation of pipelines and prohibiting issuance of fraudulent oil certificates. This "was allowed to pass" the legislature in a revised, weakened version, and "only because it contained a clause repealing" Sherman's law of 1874 and after Patterson of the Committee on Legislation had struck out the provision for payment of court-appointed examiners. The examiners were thus left at liberty to accept any bribe. The revision also relieved the pipeline company officials who transgressed this new act from the earlier penalties of fine and imprisonment. Pipeline regulation, therefore, was weakened rather than strengthened and extended as Sherman intended. The General Council's general pipeline bill, with provisions for chartering companies with the right of eminent domain, was defeated at Harris-

[6] Commonwealth of Pennsylvania *v.* Pennsylvania Railroad Company, *Printed Testimony,* pp. 540–543, 581–583, 590, 602–605, 610, testimony of B. B. Campbell; *History of the General Council of the Petroleum Producers Association* (Titusville, 1880), pp. 6–9.

[7] *History of the General Council,* p. 9; Sherman Papers, *Pamphlets,* Commonwealth of Massachusetts, *General Railroad Laws* (Boston, 1878), *passim* for notations.

[8] *History of the General Council,* p. 9.

burg.[9] As chief counsel Sherman undoubtedly had a share in drafting that bill also.

As a member of the Titusville Producers Union and its delegate to the General Council he shared the Committee on Legislation's reported disillusionment of May 15, 1878:

"How well we have succeeded at Harrisburg you all know. It would be in vain for your committee to describe the efforts of the council in this direction. It has been simply a history of failure and disgrace. If it has taught us anything it is that our present lawmakers, as a body, are ignorant, corrupt and unprincipled; that the majority of them are directly or indirectly, under the influence and control of the very monopolies against whose acts we have been seeking relief. . . . There has been invented by the Standard Oil Company no argument or assertion, however false or ridiculous, which has not found a man in the Pennsylvania Legislature mean enough to become its champion." [10]

During the sessions of 1878 to 1879 the Committee on Legislation experienced a second series of defeats at Harrisburg and Washington.

Sherman's labor as chief counsel included assisting the Committee on Legal Remedies. As the other avenues of expected relief for the producers were attempted without decisive results—by the committees on Transportation, Pipe Lines, Patents, and Refining—the work of the Committee on Legal Remedies rose to first importance. It began earnest work, however, before the General Council attempted in the summer of 1878 to stop the drilling of new wells and tried to induce individual producers to erect storage for their production so as to be free from enforced sales. The object was to stop the decline of the price of oil which fell to $1.125 on June 12 when the visible supply rose to 5,000,000 barrels. The unprecedented development of the McKean County field by "a large number of operators who had not before been producers" frustrated this policy.[11] The short-lived Equitable and the belated Tide Water

[9] *Ibid.;* Sherman Papers, Sherman to C. W. Stone, March 9, 1878, Stone to Sherman, March 11, 1878.

[10] *History of the General Council,* pp. 9–10.

[11] *Ibid.,* pp. 7–8; *Titusville Morning Herald,* June 12, 1878.

pipelines provided but limited relief to producers eager for independent pipeline facilities.

On March 6, 1878, the Executive Committee of the General Council asked Sherman for a written opinion:

Have the Producers, individually or collectively, any redress, in any Court of Equity under the laws existing at this date, for loss or damage to their Business by reason of the combination entered into between the Standard Oil Co & the Railroads, which has through an unlawful and undue discrimination in Freights, made them the only Buyer of our product, to our great detriment and loss? [12]

He replied that "*collectively*" the producers had no such remedy unless they were lawfully organized "in a co-partnership or as a corporation," which "would have a right of action for such unlawful combination, for damages sustained *subsequent* to the organization; but not for wrongs done to individuals prior thereto." Each producer who could prove individually that he had been "actually damaged in his business, *has* a right of action" against a proven unlawful combination formed to profit from those excluded from it. The General Council promptly adopted a plan for "Legal Organization of the local Unions." They approved and levied an assessment upon their members to finance legal expenses. In Titusville this produced the Titusville Petroleum Producers Association, Ltd., of which Sherman was a shareholder.[13] He billed it for the balance of his retaining fee as counsel of the General Council and for his legal services.[14]

In May the Committee on Legal Remedies advised that the General Council retain counsel to enforce existing laws prohibiting "conspiracies in restraint of trade and discrimination in freight rates by common carriers." That body authorized Sherman to take preliminary steps to develop this method of relief for the producers whose situation was rapidly deteriorating. In view of the legislature's failure to provide remedial legislation Governor J. F. Har-

[12] Sherman Papers, Henry Byrom to Sherman, March 6, 1878.

[13] *Ibid.*, Sherman to Byrom, March 13, 1878, T. C. Joy to Sherman, March 15, April 5, 1878.

[14] *Ibid.*, Sherman to Titusville Petroleum Producers Association, Ltd., April 30, 1878.

tranft suggested that it appoint a commission to study the oil industry's problems and recommend "legislation acceptable to all classes and just to all interests."[15]

Then, in early June, when the producers' tanks were running over, the United Pipe Lines refused to buy oil except for immediate shipment at below the market price. This announcement precipitated such anger, excitement, and danger of mob violence that President Campbell telegraphed John Scott, President of the Allegheny Valley Railroad, saying ominously that "unless immediate relief is given and the immediate shipment plan abandoned" there would occur "excess which the more prudent part of the people cannot prevent, and the scenes of last July in Pittsburgh" would be repeated.[16]

Conferences ensued with a Standard Oil representative, who proposed that the producers combine with Rockefeller, who would pay them a price for crude oil "based upon the relative prices of refined."[17] Although this appeared "superficially pleasing," when the Committee on Legal Remedies "discovered that acceptance would involve surrender of the principle that such monopolies were contrary to law and policy, and that in the only two known instances where producers had been induced to enter into similar contracts they had been violated by the Standard party whenever they were likely to be profitable to the producer, the Council declined to confer with John D. Rockefeller upon the subject." After an informal group of producers did so fruitlessly, the Council dismissed its three-man committee on Legal Remedies and replaced it with a committee of five.[18]

In April Sherman had been associated with George Shiras, Jr., a noted Republican attorney of Pittsburgh, in a suit against the Pennsylvania Railroad.[19] Sherman was shaping plans for criminal prose-

[15] *History of the General Council,* p. 10; Pennsylvania Archives, Fourth Series, *Papers of the Governors,* IX, 650–651.

[16] Commonwealth of Pennsylvania *v.* Pennsylvania Railroad Company, *Printed Testimony,* p. 540.

[17] *History of the General Council,* pp. 12–13.

[18] *Ibid.*

[19] Sherman Papers, Shiras to Sherman, April 29, 1878.

cution of officials of the Standard Oil and United Pipe Lines for perpetuating "the immediate shipment swindle." In late June he conferred secretly with Attorney General George Lear in the producers' behalf.[20] In July he consulted Mark W. Acheson, another prominent Republican attorney of Pittsburgh, on the effectiveness of *mandamus* procedures as a means of securing the producers' rights. Sherman advised the Equitable on how owners of oil piped via its lines from Bradford to Corryville could secure the use of idle tank cars on the Buffalo & McKean and the Buffalo, New York & Pittsburgh railroads.[21] He also consulted George A. Jenks, an Oil Region attorney, while shaping the preliminary legal plans of the General Council.

Completion of the Tide Water to Williamsport was as yet nearly a year distant. The railroad trunklines were continuing to grant the Standard Oil and its subsidiaries virtually exclusive use of their facilities. The bitter discriminations of the United Pipe Lines against its producer opponents during the months of "immediate shipment," the railroads' refusal to supply tank cars to independent oil buyers and refiners who were willing to pay higher net freight rates than those charged the Standard Oil, and the carriers' initial attempts to block the Equitable's route to the Erie Canal confirmed the independent oil men's conviction that they were in the merciless grasp of the monopoly that the Rockefellers had created with the decisive aid of the great carriers.

Sherman had urged repeatedly upon the Titusville leaders of the movement "the necessity of action upon a definite plan, but nothing was done." In early July Campbell informed him that a committee of twenty-five would interview the Governor and present "a case to the Attorney General." Sherman urged that a statement of facts to inform the Governor be prepared, and petitions against certain corporations be circulated for the Attorney General's use supple-

[20] *Ibid.*, Sherman to George A. Jenks, March 24, 1878, Lyman D. Gilbert, Deputy Attorney General, to Sherman, June 25, 1878.

[21] *Ibid.*, Acheson to Sherman, July 13, 1878, Samuel G. Thompson to Sherman, July 13, 1878, Sherman to Equitable Petroleum Company, July 1, 1878.

mented by "competent *evidence* of the facts alleged, so that we could be prepared to sustain ourselves in court" in any "instituted proceedings." However, since no evidence was adduced he drafted no papers. Consequently, he declined to go with the committee to Harrisburg despite Campbell's urgent solicitation, because he believed that that figure and a few others could accomplish all that could be won by interviewing the Governor.[22]

He advised Campbell bluntly also that nothing could be "gained by . . . further expenditure of time and money in mere public demonstrations, addresses, resolutions, or interviews with high officials. I consider all that has been done in the way of procuring legislation as so much wasted time and money. Our people have neither time or money to fritter away in useless endeavors to make legislatures and officials do their simple duty." He reiterated the opinion that he had presented previously to the Council after consulting "able and learned lawyers." This was that there was only one way to cope with "individuals and corporations who, presuming upon their power are boldly acting in defiance of law and challenging you to a legal contest if you consider yourself injured." That was to "prosecute them criminally and civilly. I have all the confidence in the courts. In Pennsylvania they are uncorrupted, and I believe incorruptible. The law is plain. The Courts are just. You have been wronged. Let the law protect you." For such a case "*evidence* that is pertinent and competent" and "*Money* to pay the legal expenses" were indispensable. His "learned" legal advisers agreed. They were willing to proceed "with a certain enthusiasm which is not always aroused, and for which the lawyers as a class get little credit." This was not a matter of fees. "Lawyers," he said, "who properly understand the high responsibilities and duties of their profession, are glad to embark in a litigation that has for its object the establishment of great principles, the vindication of the right, and the maintenance of the Supremacy of the law." He urged Campbell "to lose no time in inaugurating the necessary legal proceedings, upon a

[22] *Ibid.,* Campbell to Sherman, July 25, 29 (two telegrams), 1878, Sherman to Campbell, July 29, 30, 1878.

definite and well-considered plan, and fortified by the necessary means to secure its success." [23]

As he penned this recommendation Campbell's rump committee persuaded the Governor and the Attorney General on July 30 to intervene in the crisis of the oil trade by executive-directed litigation. Governor Hartranft requested that the aggrieved producers present to him a detailed memorial on the situation setting forth their wrongs on the basis of which Lear could act, a document such as Sherman had urged that Campbell take with him. Accordingly, Sherman drafted for the General Council, while communicating with Campbell in code, the famous *Appeal to the Executive of Pennsylvania.* After it was signed by Campbell and thirteen district representatives of the Producers Unions Sherman presented it to Hartranft and Lear on August 15.[24] In the interim he had handed to Secretary Byrom his recommendation that (1) criminal conspiracy indictments be secured against United Pipe Lines' managers who depressed the oil market by well-known methods, (2) *quo warranto* proceedings be instituted against railroad and pipeline companies refusing to perform or who were violating their common-law obligations, and (3) civil damage suits be instituted in behalf of each individual or company injured by both wrongs. "The cheat called 'Immediate Shipment' is a transparent fraud," he added, "and I do not see how it would be possible for the United Pipe Line Co. to avoid the rendering of judgment against it in each case, for heavy damages." Each "has his remedy," whether as an individual or limited partnership association. The Unions' limited partnership associations should "tender" their oil, demand its receipt and shipment, and in the name of the association bring suit for damages, thus "lessening the expense to individuals and giving strength to the movement." Thus, through these local business branches of the

[23] *Ibid.,* Sherman to Campbell, July 30, 1878.

[24] *Ibid.,* Campbell to Sherman, August 1, 1878, Campbell to Patterson, August 6, 22, 1878, Sherman to R. B. Brown, August 22, 1878, Sherman to B. Whitman, September 2, 1878, George Lear to Sherman, October 3, 1878; Commonwealth *v.* Pennsylvania Railroad, *Printed Testimony,* p. 583; *History of the General Council,* pp. 13–14.

Producers Unions Sherman wished to institute litigation paralleling whatever action the state administration could be induced to undertake.[25] The situation confronting the independent oil men was worsening daily.

*The Appeal to the Executive of Pennsylvania,* which Sherman presented to Governor Hartranft in a private conference, persuaded him to adopt the General Council's strategy. The document was an antimonopolist manifesto of great significance and invoked the Commonwealth's obligation to come to the relief of citizens who were seriously injured in their businesses by a "foreign" monopolistic corporation. It contrasted the prosperous, competitive oil industry of 1871 with the depressed, monopoly-controlled industry of 1878. The trunkline railroads' illicit discriminations in the Standard Oil's favor were a double betrayal of the Commonwealth by its chartered corporations. The carriers' executives had become shortsighted, craven, lawless allies of grasping, ruthless monopolists. In calm prose the *Appeal* invoked the stereotypes produced by five years of bitter business conflict in the Oil Region. It petitioned Governor Hartranft to restore free enterprise in the petroleum industry by prosecuting the lawless railroad barons and Rockefeller coterie.

A vivid portrayal of specific injuries inflicted on the producers pointed up a review of the South Improvement Company conspiracy and the Standard Oil's consistent enjoyment of railroad favoritism. "Immediate shipment" was depicted as the most extraordinary example of Rockefeller vindictiveness and maladministration. *The Appeal* stated that the Standard Oil had been enabled to realize the objectives of the "conspiracy" by the Pennsylvania Railroad's defeat of the "Free Pipe Law" in the legislature, by the trunkline pool's carrying of oil from the pooled pipelines 22¢ per barrel cheaper than that from independent pipelines, by the Lake Shore Railroad's minority holding in the United Pipe Lines Company, and by the victory of Ohio and New York railroad capitalists over the Pennsylvania Railroad in the 1877 oil rate war. The Standard Oil's defiance

[25] Sherman Papers, Sherman to Byrom, August 14, 1878.

of the congressional investigation of 1876 had defeated the appeal of "the people of Pennsylvania" to Congress. The same "covert" influences and "familiar methods" of the "legislative agents" of great interests had defeated the state antidiscrimination bill at Harrisburg, although its enactment would have rebuilt Pennsylvania's refineries and Philadelphia's oil export trade and freed northwest Pennsylvania from vassalage to a "foreign corporation."

The people who had chartered railroad corporations as "public servants" and "common carriers," instead of receiving equitable service from them were "now on the defensive, their rights denied . . . and they are challenged to enter the Courts to establish them." Meanwhile, their denial wreaked "irreparable injury to their business" and deprived the independents of transportation.[26] The United Pipe Lines' duplicity, coercion of producers, use of their tanks for its storage of oil, delays in payment and piping, its "overissue" of certificates, and "immediate shipment swindle" left "the People" but "two remedies" for the oppression of "the Pipe Line and Railroad Companies . . . an appeal for protection, first to the Law of the Land, next, to the higher law of Nature!" *The Appeal* thundered threateningly. It castigated "the Autocrat of a foreign corporation" who had destroyed most of the Oil Region refineries, eliminated "competitive buyers," and subjected the producers to "a dictated price." *The Appeal* then asked Governor Hartranft to act immediately "to compel obedience to law, and the performance by chartered companies of their public duties." Republished widely *The Appeal* confirmed inherited Jacksonian stereotypes and mobilized a public opinion essential for effective executive action.[27]

This public opinion had been created in the Oil Region by the Producers Unions' energetic organizing campaign, which reached a climax after presentation of *The Appeal* in a monster demonstration at Bradford.[28] The strengthened Producers Unions provided the

[26] Cf. *History of the General Council,* pp. 10–13.

[27] *An Appeal to the Executive of Pennsylvania* [Titusville, 1878], pp. 3–11; *Titusville Morning Herald,* August 30, 1878.

[28] *Titusville Morning Herald,* June 14, August 14, 15, 1878.

Governor with the support of a powerful, organized business interest.

Hartranft believed in the necessity of national regulation of interstate commerce. He believed that it was to the interest of Pennsylvania and the railroad stockholders that all parts of the Commonwealth enjoy "free and fair transportation." If secured, this would remove "the bitter and growing prejudice" against the railroads by eliminating its causes. He took immediate advantage of *The Appeal* to request formally that the Pennsylvania Railroad observe its legal obligations to shippers. When Thomas A. Scott refused to pledge compliance, Hartranft conferred with Roger Sherman as the counsel of the Producers Union at Doylestown on August 17 in the presence of Attorney General Lear and Assistant Attorney General Lyman D. Gilbert. It was decided there to utilize *The Appeal* "to secure from the Supreme Court of Pennsylvania an authoritative and binding declaration . . . of the duties and obligations of corporations." [29] Article XVII, Sections 3 and 7 of the Pennsylvania Constitution, with its assertion of the "equal right" of shippers on railroads without "undue or unreasonable discrimination" and prohibition of drawbacks or rebates—which Sherman's partner Beebe had helped to insert in 1873—provided the basis for this action. That these Sections had not been implemented by appropriate legislation was due to the opposition of the Pennsylvania and Reading railroads and their business favorites. The humiliating defeats inflicted upon the Legislation Committee of the General Council at Harrisburg had already taught Campbell and Sherman that both branches of the state legislature were packed with members biased in favor of the Pennsylvania and Reading railroad interests. During his successful negotiation with Governor Hartranft, Sherman persuaded him to resort to *quo warranto* proceedings.[30]

Since the Attorney General lacked the staff and a budget adequate

[29] Pennsylvania Archives, Fourth Series, *Papers of Governors,* IX, 693–694.

[30] Commonwealth *v.* Pennsylvania Railroad, *Printed Testimony,* pp. 581–583.

for conducting the litigation involved in this appeal to the Supreme Court, Hartranft and Lear entrusted its conduct, to Campbell's great disgust, to the private counsel employed by the petitioners, subject to the direction of the Attorney General.[31]

After his return to Titusville Sherman reported that the conference was "very satisfactory" and that Governor Hartranft was "sincere in his desire to make all Corporations discharge their duties." Lear had instructed him to notify the United Pipe Lines that he would move in the Venango County Common Pleas court for a writ of *quo warranto* the following Monday. Sherman asked Lear what would be "the next step" should that company consent immediately or be compelled to perform its "corporate duties" and the railroads "refuse as heretofore to furnish cars to carry petroleum" delivered by it to them. Lear instructed him in such a contingency to proceed against the offending railroads "in the same way," but remarked that "all the principles we contend for could be established by the proceedings against the United Company."[32] On his recommendation the Committee on Legal Remedies named the counsel to conduct the litigation. Sherman was placed in charge. Associated with him were Shiras, Acheson, Jenks, and subsequently E. Copee Mitchell of Philadelphia. Sherman worked closely with them while maintaining liaison with the Governor and Attorney General.[33]

Campbell, meanwhile, began gathering funds to finance the litigation. The Unions appointed collectors for every district, to whom many persons of limited means subscribed small amounts during the peak of enthusiasm aroused by the prospect of success in the courts, in supplement to larger contributions from leading producers and certain "outside" New York oil buyers and refiners. The latter, incidentally, pledged only the small sum of $500. The producers bore the real burden of financing the suits that ensued but

[31] Pennsylvania Archives, Fourth Series, *Papers of Governors,* IX, 773–774, Governor Henry M. Hoyt's Biennial Message to the Assembly, "Freight Discrimination," January 4, 1881; Commonwealth *v.* Pennsylvania Railroad, *Printed Testimony,* pp. 575–577, 579, 581–584.

[32] Sherman Papers, Sherman to R. B. Brown, August 22, 1878.

[33] *History of the General Council,* p. 14.

contributed in actual cash an insufficient sum for fully effective action. The cooperation with the New York City independents derived from the joint refusal of the United Pipe Lines and the Allegheny Valley Railroad to ship oil to New York except from Bradford, which deprived New York independent refiners of indispensable Allegheny Valley oil to mix with the Bradford crude. Henry C. Ohlen, the leading independent New York buyer, bluntly informed Campbell that the independent producers' market in the metropolis would vanish if they submitted to this refusal, and pledged the modest financial aid.[34]

President Campbell was overjoyed. He was equally pleased when S. S. Avery of H. L. Taylor & Company joined in opposition to the Allegheny Valley Railroad's refusal of cars for Butler oil for shipment eastward while providing cars for its shipment to the Standard Oil-controlled Pittsburgh market.[35] After this Campbell and Sherman felt justified in believing that they held the game in their hands. After all, they enjoyed among the producers the support of some influential Republican legislators such as Lewis Emery, Jr., the support of the powerful *Morning Herald,* and the association of Shiras and Acheson who as leading Pittsburgh Republicans had great influence with the state administration.

There was an immediate danger that the indignant and excited producers and general public in the Oil Region would take the law into their hands as "immediate shipment" continued while great quantities of oil flowed to waste on the ground as the United Pipe Lines refused to pipe it to the railroads. Great crowds of producers lined up daily before the offices of the Standard Oil and United Pipe Lines in Bradford that late summer in attempts to sell their oil at a heavy discount for "immediate shipment." They voiced their resentment in spectacular mass meetings, as the railroads refused to supply tank cars to the few independent pipelines.

[34] Commonwealth *v.* Pennsylvania Railroad, *Printed Testimony,* pp. 577–579; Sherman Papers, H. C. Ohlen to W. H. Nicholson, August 21, 1878, Nicholson to B. B. Campbell, August 22, 1878.

[35] Sherman Papers, Campbell to Patterson, August 22, 1878, Avery to Campbell, August 23, 1878.

Meanwhile, Sherman conferred frequently with Campbell, Patterson, other leading producers, and his associate counsel in an intensive attempt to cope with the climax of the Standard Oil's efforts to stamp out the remaining business competition in the oil industry. Campbell and Sherman also received detailed data on the practices of the United Pipe Lines from producers and other businessmen. The aggressive, individualistic President of the General Council was simpler, less astute, than Sherman with whom he worked somewhat erratically, not always keeping his promises as to policy and procedure punctually or completely. Sometimes Campbell accepted statements from the accused carriers in the Standard Oil combination that data supplied to Sherman proved to be false. The sturdy, consistent antimonopolist Patterson, third member of the trio who planned the General Council's high strategy, was the least astute of its members. A courageous extrovert, he had accepted a moderate fee from the United Pipe Lines—although he was not a lawyer—when he assisted in the enforcement of Sherman's pipeline act of 1874, unaware that this laid him open to the charge that he fought antimonopolist battles for private profit. In late August Patterson decided upon publication of *The Appeal to the Executive of Pennsylvania* and so informed Hartranft. Insight into the relationships within the trio is provided by a telegram to Patterson from Campbell on September 3, 1878: "Meet me at Sherman's office on arrival of train this evening." [36]

Immediate violence on August 23 by a huge mass meeting at Bradford was prevented only by Sherman's telegraphed promise in the name of the Commonwealth that it would move immediately for the relief of the Bradford producers and require the United Pipe Lines to "perform its duties or cease to exist." [37] Three days later he aided Assistant Attorney General Gilbert at Franklin in filing be-

[36] *Ibid.,* Patterson to Governor Hartranft, August 28, 1878.

[37] *Ibid.,* A. C. Hawkins to Sherman, August 23, 1878, Sherman to Hawkins, August 23, 1878. When Sherman's telegram was read, the turbulent Bradford meeting applauded Governor Hartranft and pledged those present to support the General Council's program of enforcing the law against the oil combination (*History of the General Council,* pp. 14–15).

fore the Court of Common Pleas the Commonwealth's declaration that secured a writ of *quo warranto* against the United Pipe Lines Company that was returnable September 11. The writ required it to show cause why its charter should not be forfeited on account of its "unjust and oppressive discriminations" in violation of it. The United Pipe Lines' repeated demurrers were rejected on September 26 when Judge Charles E. Taylor gave it twenty days in which to reply. Then, when the reply was filed, he scheduled a jury trial for December 1.[38]

Both Campbell and Sherman recommended that *quo warranto* proceedings be instituted against the Pennsylvania Railroad Company also after President Scott replied in a newspaper interview to *The Appeal,* declaring: "We tried the system to which the men who are making the complaints against the Standard's contracts wish us to go back and found it did not work well." Unfortunately, Campbell exhibited a lack of consistency in recommending to Governor Hartranft that he convoke a special legislative session to enact legislation to enforce the Constitution's provisions guaranteeing equal treatment of shippers on common carriers, because it was doubtful that "our relief can come through the Courts, as the law now stands" because "our rich & powerful opponents" would resort to every possible device to delay a decision. This Campbell recommended, despite the agreed-upon policy to take the great issues to the Pennsylvania Supreme Court and in the face of the General Council's defeat in the legislature of the preceding spring, and when but a few months were left of the Hartranft administration.[39] Such a lack of acumen hardly inspired the Governor or Sherman with respect for Campbell's political realism.

On that very day, August 31, Sherman was preparing to widen the scope of the Commonwealth's *quo warranto* suits to include the

[38] Sherman Papers, Gilbert to Sherman, August 24, 1878, Sherman to Gilbert, August 24, 1878, Sherman to Emery, Jr., September 6, 1878, J. H. Caldwell to Sherman, September 5, 1878; *Titusville Morning Herald,* September 12, October 3, 1878.

[39] Sherman Papers, Campbell to Sherman and enclosure (B. B. Campbell to Governor John F. Hartranft, August 31, 1878), September 1, 1878.

Standard Oil Company of Pennsylvania,[40] an action which was not instituted. At that time he was acting for the General Council against the United Pipe Lines in several counties in local litigation that sprang from its refusal to take oil which it had "capacity to carry." [41] These suits were analogous to cavalry engagements preliminary to the main clash of contending armies such as he had participated in during the Civil War. Lear acquiesced in his request that injunction proceedings be instituted to restrain the United Pipe Lines from the abuses complained of and instructed Sherman to draw up the bill.[42]

On Lear's instruction, after consultation with Campbell and Patterson, Sherman prepared injunction bills against the Pennsylvania Railroad Company, Atlantic and Great Western Railroad Company, Dunkirk, Allegheny & Pittsburgh Railroad Company, and Lake Shore & Michigan Southern Railroad Company. These charged the railroads with conspiracy with each other and with the Standard Oil to monopolize the oil business. He drafted another bill directed at the United Pipe Lines. These bills he sent to Lear for criticism and received them back approved and with instructions for filing in the Supreme Court, Western District. This was a procedure that the Attorney General had preferred from the beginning since by this means it was possible to secure an early decision upon the great issues at stake.[43]

Campbell asked Shiras early that September if a *quo warranto* against the Pennsylvania Railroad should not also be instituted. Shiras informed Sherman that Hartranft vetoed it. Instead Shiras recommended strongly trying "the effect of an indictment" since the "managers of the Standard combination and of the Penna R R Co are guilty of conspiracy," and judges and juries in any northwest

[40] *Ibid.,* Sherman to H. A. Miller, August 31, 1878.

[41] *Ibid.,* Sherman to Wm. McNair, August 31, 1878.

[42] *Ibid.,* Sherman to Gilbert, September 2, 1878, Lear to Sherman, September 4, 1878, Sherman to Lear, September 6, 10, 1878.

[43] *Ibid.,* Sherman to Lear, September 10, 1878, Lear to Sherman, September 14, 1878, Campbell to Sherman, September 26, 1878; Commonwealth *v.* Pennsylvania Railroad, *Printed Testimony,* pp. 575–577; *History of the General Council,* pp. 13–14.

county would give "justice against them." He deferred to Sherman's judgment as to whether or not the facts could be adduced adequately to secure conviction. Sherman agreed and was anxious to institute such a prosecution.[44]

Hurried consultations with Hartranft, Lear, and the associate Commonwealth counsel, when the United Pipe Lines sought to transfer its Bradford business back to its subsidiary, the American Transfer Company (a New York corporation), and the launching of five injunction suits against the railroads and the United Pipe Lines pushed this plan aside. Sherman also briefed leading producers on the litigation in advance of securing formal approval of it from the General Council. At Bloss's request he wrote and telegraphed to the *Morning Herald* an account of the filing of the injunction bills and of the test that that procedure presented to the Pennsylvania Railroad. The injunction bill against the Pennsylvania Railroad, signed by Lear, Gilbert, and the producers' counsel, was published there in full.[45]

Immediately after the United Pipe Lines abandoned "immediate shipment" the Titusville Petroleum Producers Union, directed by the General Council, petitioned for an investigation of the petroleum business by General William H. McCandless, Secretary of Internal Affairs of Pennsylvania. The statute of May 11, 1874, directed him to "exercise a watchful supervision over the railroad, banking, mining, manufacturing and other business corporations of the State, and to see that they confine themselves strictly within their corporate limits." He was obliged to investigate any sworn charge that a corporation exceeded its powers by infringing upon

[44] Sherman Papers, Shiras to Sherman, September 11, 1878, Sherman to Shiras, September 12, 1878.

[45] *Ibid.,* Campbell to Sherman, September 14, 1878, Sherman to Gilbert, September 16, 1878, Lear to Sherman, September 26, October 3, 1878, Sherman to Lear, September 21, 28, 1878, Campbell to Sherman, September 26, 1878, Gilbert to Sherman, October 1, 1878, "P" to Sherman, September 26, 1878, Shiras to Sherman, September 26, 1878, Sherman to Shiras, September 20, 30, 1878, H. C. Bloss to Sherman, October 7, 1878; *Titusville Morning Herald,* October 9, 10, 1878.

individual rights and to report his findings to the Attorney General, who should redress any substantiated grievances by legal action. Cooperating with Sherman, who had drafted the petition, President Campbell wrote McCandless invoking this act with a sworn charge against the Standard Oil combination. McCandless instructed his deputies, J. Simpson Africa and James Atwell, to investigate. When Africa came to Titusville on September 19 to hold hearings, Sherman handed him in the rooms of the local Producers Union a detailed petition stating their alleged wrongs.

Africa took testimony there, at Pittsburgh, and at Harrisburg until October 8. This established the facts of a monopolistic oil combination benefiting from a system of rebates from transportation companies, the illegal acts of the United Pipe Lines, and the refusal of the railroads to supply tank cars to independent oil men. During the hearing the Standard Oil and railroads developed "the line of policy they intended to pursue—deliberate and defiant contempt of the law and its processes," Sherman observed later. This occurred despite the United Pipe Lines' private allegations that it had acted legally under the charter of the American Transfer Company in Venango County.

Officials of these corporations refused to respond to Africa's subpoenas to appear and testify or to communicate with him directly. He was obliged to report that he had been unable to secure "definite knowledge of the details of the conspiracy." [46] Atwell did take ninety-two pages of testimony. Ignoring this and while absent from Harrisburg, McCandless made his "report" to Lear by telegraph, asserting that "no cause had been presented to him beyond the ordinary province of individual redress." The producers greeted this with derision and hung McCandless in effigy. The stunned Sherman observed that opinions differed as to whether the Secretary "was afflicted with the same kind of obtuseness that shadowed the Pennsylvania Legislature" when the producers' antidiscrimination

[46] Sherman Papers, Africa to Sherman, September 5, 1878; Commonwealth *v.* Pennsylvania Railroad, *Printed Testimony,* p. 576–577; *History of the General Council,* pp. 15–16.

bill was beaten "or as to whether there was a consideration paid for this performance." [47]

Meanwhile, the *quo warranto* suit against the United Pipe Lines made Governor Hartranft the hero of the hour in the Oil Region. The political credit for this redounded to the credit of the Republican party, since McCandless was a Democrat! However, all three gubernatorial candidates during the campaign of 1878 pledged to continue Hartranft's policy to bring the railroads and their fostered monopolies to book. General Henry M. Hoyt of Wilkesbarre, the Republican nominee, ran on an antimonopolist, antidiscrimination platform. However, when he came to Titusville on September 7 he was escorted on a carriage tour of the city and its environs by John D. Archbold of the Standard Oil and David McKelvy of the Seaboard Pipe Line Company. Thus, he was confronted under the late summer sun with the power of the oil monopoly and the might of its opposing independent business and popular antagonists. Archbold and McKelvy presented to him their views of the "oil question" as he praised the city's beauty.[48]

Through Benjamin Whitman, Sherman pressured Hoyt's Democratic opponent to declare publicly in favor of enforcing the constitution and laws relative to corporations. Sherman's personal conviction, forwarded to him by Whitman, was that "the next great issues in Pennsylvania will arise from a determination by the people that corporations, great and small, shall be held to a rigid performance of their corporate functions, and strict accountability to the state." The evils to be remedied, he added, prevailed in the entire state and throughout the North. "You know that Pennsylvania is a Ring-ridden State," he said, where candidates of the major parties were taken from the dominant political "ring, beared up and put forward by it. The Ring must be broken." However, although he was a Democrat, Sherman insisted that the producer-financed

[47] *History of the General Council,* pp. 16–17; Sherman Papers, James Atwell to E. G. Patterson, September 13, 1878, Atwell to R. B. Brown, September 13, 1878, Sherman to Patterson, September 24, 1878; *Titusville Morning Herald,* September 20, 21, October 7, 1878.

[48] *Titusville Morning Herald,* August 28, September 9, 1878.

Commonwealth Suits "should be removed as much as possible from *political* discussion." They were not political in character, as he informed Lear when the Philadelphia *Times* impugned his and Hartranft's motives in October. He hoped that Pennsylvania would be freed from ring rule by bipartisan action.[49]

Hoyt's resounding victory in November was regarded in the Oil Region as a victory for the independents, for antidiscrimination, for competitive free enterprise. The General Council of the Producers Unions expected confidently that he would continue unflinchingly Hartranft's program of court action against the Standard Oil and its railroad confederates. However, the Republican party had been interpenetrated some time since by the powerful influence of the Pennsylvania Railroad and of the Standard Oil with which it was openly allied. This the now complacent Campbell should have remembered.[50]

[49] *Ibid.*, October 7, 1878; Sherman Papers, Sherman to Whitman, September 2, 1878, Sherman to Brown, August 31, 1878, Lear to Sherman, October 3, 1878, Sherman to Lear, October 9, 1878; Commonwealth *v.* Pennsylvania Railroad, *Printed Testimony*, p. 588; *History of the General Council*, pp. 17–21.

[50] *History of the General Council*, pp. 17–21; *Titusville Morning Herald*, October 7, 1878.

# CHAPTER VI

# *Exposing the Standard Oil Combination*

ROGER SHERMAN achieved his greatest public distinction as prosecutor of the Commonwealth Suits during the years from 1878 to 1880. In this role he made the most formidable attempt in the courts to overthrow the "new type of monopoly" that Rockefeller had created and which the Benson group were to weaken by their successful Tide Water Pipe Company, Ltd. The new-type monopoly presented American business with its most serious problem during the prolonged "Gilded Age." Sherman's great objective was the rehabilitation of the competitive system in the petroleum industry. To the independent oil men and the Commonwealth of Pennsylvania which retained him this was the only desirable economy. They knew that it was justified by the legal and constitutional system that reached back to Tudor England.[1]

Sherman was attempting with the support of Governor Hartranft to enforce this system of legal restraints upon the railroad allies and chief pipeline subsidiary of the Standard Oil which defiantly had created a monopolistic business empire in the midst of the free society. Sherman wished to prosecute the railroad and pipeline executives and those of the Standard Oil criminally for conspiracy. The Standard Oil combination rested, he and Hartranft claimed, upon the perversion of the common carrier functions of the pipelines and railways. The correction of this situation was the great objective of the Suits. Success in the litigation, they believed, would destroy the chief foundation of the oil monopoly.

Sherman was occupied for some months with courtroom prelimi-

[1] Nevins, *Study in Power,* I, 288, 303.

naries, during which the attorneys of the United Pipe Lines Company and railroads interjected every possible obstruction. They attempted to obfuscate the issues so as to confuse the courts and public, whose informed indignation supported the producers. The limited financial resources of the Petroleum Producers Unions made this tactic of delay effective.

Sherman consulted President Campbell of the General Council and his associate counsel frequently on the strategy of the litigation. Unlike Campbell, who demanded "boldness & decision" lest "we are sold out," he did not "fear the Supreme Court" nor impute to it the possibility of "a gross denial of justice." He led in the preliminary maneuvers that persuaded the Supreme Court to accept jurisdiction in the injunction proceedings against the four railroad defendants and United Pipe Lines. An information, supported by affidavits based upon the testimony of Henry C. Ohlen and Peter Lombard before Atwell, drew blanket denials from the defendants. Sherman and his associates then withdrew their motions for preliminary injunctions, applied instead for "relief on final hearing," and persuaded the court to refer these suits, and subsequently the *quo warranto* against the United Pipe Lines (which was appealed promptly to it), to a master who would establish the record of facts for each. This was General J. Bowman Sweitzer, who was appointed on November 26.[2]

Meanwhile, Sherman had crossed swords with the United Pipe Lines' attorneys led by S. C. T. Dodd before Judge Taylor at Franklin in the initial *quo warranto* proceeding. There the Titusvillian presented a masterly exposition of the thesis that the United Pipe Lines was a recalcitrant common carrier, described specific discriminations practiced between its customers, asked for judgment of "*ouster*," and handed Taylor his brief of authorities.[3]

[2] Sherman Papers, Sherman to Ohlen, November 4, 1878, Sherman to Shiras, October 8, 14, 1878, Campbell to Sherman, October 14, 18, November 11, 1878, L. D. Gilbert to Sherman, October 10, 1878, Sherman to S. C. T. Dodd, October 18, 1878, Shiras to Sherman, October 20, 1878; *History of the General Council,* pp. 16–17; *Titusville Morning Herald,* November 21, 1878.

[3] *Titusville Morning Herald,* November 2, 1878; Sherman Papers, Sherman to Gilbert, November 1, 1878, Sherman to Judge Taylor, November 2, 1878.

Simultaneously Sherman gathered evidence widely for the injunction proceedings. The Producers Unions supplied data on specific abuses, as did the Pittsburgh Chamber of Commerce which had long fought railroad discriminations. At Cleveland his representative interviewed Samuel Andrews, formerly of Rockefeller, Andrews, & Flagler. The affidavits gathered in preparation for argument of the abandoned motions for preliminary injunctions were of obvious value. Sherman asked Lear vainly to defray the cost of witnesses and the Master's hearings from his budget.[4]

Not all Sherman's behind-the-scenes strategy has come to light, but much can be inferred from his formal responsibility for the Suits and his position as general counsel of the General Council of the Producers Unions. He attended the General Council's monthly meetings which he informed of the progress of the Suits. He may have inspired the successful negotiations that allied the independent New York refiners and oil buyers with the General Council. It was as counsel of the Producers Unions that he induced John Scott to fulfill the Allegheny Valley Railroad's contract to lay a sidetrack for the independent Emlenton Producers Pipeline in defiance of the United Pipe Lines.[5]

As chief Commonwealth associate counsel he conferred repeatedly with General Sweitzer, with whom he developed a most cordial relationship, with regard to witnesses to be summoned, documentary evidence to be subpoenaed, and the locations of the hearings. Among those summoned to testify at the opening hearing at the Girard House in Philadelphia relative to the injunction bill against the Pennsylvania Railroad Company was its third vice president, Alexander J. Cassatt, his subordinates, and William G. Warden, the Philadelphia member of the Standard Oil combination. Subpoenaed as evidence were the Pennsylvania Railroad's freight

[4] Sherman Papers, Campbell to Sherman, October 27, 1878 and enclosure, Sherman to Lear, November 18, 1878, Sherman to Shiras, November 6 (twice), 7, 1878, Jenks to Sherman, November 7, 1878.

[5] *Ibid.,* T. C. Joy to Sherman, October 11, 1878, Sherman to John Scott, November 11, 18, 1878, Scott to Sherman, November 12, 19, 1878; *Titusville Morning Herald,* November 21, 1878.

tariffs and schedules, oil rate contracts, vouchers, and settlements with other railroads, transportation agencies, and the Standard Oil. Sherman wished to get down to the bedrock of decisive evidence immediately.[6]

The press then announced that the Standard Oil would propose negotiation of an agreement with the General Council of the Producers Unions. Sherman protested immediately to Campbell in a masterly statement of the legal implications of such a negotiation which, he declared, would "be destructive of all that has so far been contended for, and utterly futile as a means of affording relief from the burdens oppressing the industry." Any such agreement would be "void, and incapable of enforcement in any court." If the General Council now abandoned prosecution of the Suits after formally securing the cooperation of Governor Hartranft's administration, the Commonwealth's aid could never again be enlisted in the producers' behalf. Furthermore, the principles for which the Producers Unions contended "are of national importance." The "great industries of the country are destined to pass under the control of similar combinations unless the law interposes to check them." No agreement with the Standard Oil "can determine the law," Sherman urged cogently, while its executives' faithlessness made any "arrangement" with them "temporary."

After thus depicting the great importance of the Suits for American industry and law he informed Campbell bluntly that half the producers' troubles came from their disposition "to barter away present and future rights for a mess of potage, and the forgetting of their own importance when acting in concert." "At this critical time," he said, when every nerve should be strained to maintain the position the producers had "deliberately and wisely taken, I hope your members will not be misled with a delusive promise of compromise." The only acceptable settlement of the Suits must be a recorded court judgment condemning combinations between common carriers and dealers in a staple commodity "by which the latter

[6] Sherman Papers, Sherman to Shiras, December 3, 5, 1878, Shiras to Sherman, December 5, 1878.

are enabled to become rich at the expense of the laborer, the owner of the producing land, and the stockholders of the carrying companies." [7] Thus Sherman disclosed the high caliber of his legal statesmanship as he quashed the rumored negotiation.

He was snubbed by George Lear, possibly in retaliation, when the Master's hearings opened on December 10. Yielding to pressure from Thomas A. Scott, who complained that the Pennsylvania Railroad was being singled out as an especial object of attack, the Attorney General and his Assistant were pointedly absent that day, while Lear telegraphed a request to Sweitzer for a postponement and informed Shiras, not Sherman, that the "Commonwealth must be fair to both parties" and retain control of the Suits. Shiras had been informed the day before by John Scott, the Pennsylvania's General Solicitor, that the Commonwealth would not be ready to proceed as scheduled, and that Sherman and he would do well to travel to New York instead!

The obvious pairing of the Pennsylvania's and the Standard Oil's opposition to the Master's hearings became evident on December 14, after these had been postponed, when Lear explained to Shiras that Secretary McCandless's report embarrassed the administration's conduct of the Suits. Yielding to Thomas A. Scott, Lear ruled that Sweitzer's hearings should commence with the "examination of witnesses . . . who have suffered from the want of transportation and unlawful discriminations" so that a *prima facie* complainant's case might be made first as justification of the Suits "before going into the books and investigating the private affairs of the principal defendant." He insisted upon this course officially in letters to Sherman and in communications to Campbell, Elisha G. Patterson, and Lewis Emery, Jr., before the latter pair went to Washington in behalf of the "Interstate Commerce Bill." [8]

[7] *Ibid.,* Sherman to Campbell, December 7, 1878.

[8] *Ibid.,* Lear to Shiras, December 10, 14, 1878, Lear to Sweitzer, December 10, 1878, J. Scott to Shiras, December 10, 1878, Campbell to Sherman, December 11, 1878, Patterson to Sherman, December 12, 14, 1878, Shiras to Sherman, December 19, 1878, Sherman to Lear, December 26, 1878, Lear to Sherman, December 26, 1878; *History of the General Council,* pp. 21–22.

Sherman regarded this instruction as "preposterous and indefensible," saying that it would "operate as a perpetual injunction upon the progress of the suits, and in the end as a confession of judgment in favor of the defendants." [9] In accord with equity practice he had expected to establish the factual basis of the prosecution's case in each Suit primarily from the defendants' records and testimony of their officials. Lear's order required a complete change in the Commonwealth's tactics and multiplied the difficulties that beset the prosecution. As senior associate counsel Sherman did not resign, as Lear may have wished, but collaborated closely with Shiras in preparation for presenting the Commonwealth's evidence in each Suit. Convinced that they were being given a classic political double cross by the Cameron Republican organization they decided to continue the fight. This Hartranft encouraged them to do. Lear reassured them when he informed Sherman in an interview that after the complainant's testimony was presented he could subpoena the officials and books of the defendant corporations.[10]

Sweitzer's opening hearing of the complainant's testimony was scheduled for January 7, 1879, in the rooms of the Titusville Petroleum Producers Union. Sherman informed Campbell that this hearing necessitated presentation of "the best possible evidence" from nondefendant sources so as "to avoid a speedy defeat, for which your association, or its council would not be responsible, but which will result in the loss of all the ground you have gained, beyond the possibility of recovery." Campbell should make "every exertion to procure this evidence" immediately and provide funds to defray the "current expenses of procuring witnesses" and "the fees of stenographers," and some of the Master's fees.[11]

Receiving this letter while in Philadelphia with J. D. Benson,

[9] Sherman Papers, Sherman to Shiras, December 18, 1878.

[10] *Ibid.,* Sherman to Jenks, December 23, 1878, Lear to Shiras, December 21, 1878, Shiras to Sherman, December 23, 1878, Gilbert to Campbell, December 20, 1878, Campbell to Sherman, December 20, 1878; *History of the General Council,* p. 22.

[11] *History of the General Council,* p. 22; Sherman Papers, Sweitzer to Sherman, December 30 (twice) 1878, Sherman to Campbell, December 31, 1878.

Campbell called upon President Gowen of the Reading Railroad with whom they were negotiating the Tide Water's trans-Allegheny pipeline project. At Campbell's suggestion Gowen ordered his Comptroller to show him the Pennsylvania Railroad's monthly statements to the Reading which detailed the gross and net rates on through traffic that the latter shared and the division of oil freight receipts between the cooperating carriers. Campbell promptly subpoenaed the Reading's Assistant Comptroller to appear at the Titusville hearing and bring these papers. This was a great coup, since it enabled Sherman to establish the fact of oil freight-rate rebate payments, evidence of which he and Shiras had expected to be able to extract only from the Pennsylvania Railroad! Such was the reply of the alliance between the independent oil men and Gowen to Thomas A. Scott and George Lear.

At Harrisburg Lear stated to Campbell that he was much gratified at this development and conceded verbally that Sherman might call the official of any defendant including the Standard Oil and the United Pipe Lines except the Pennsylvania Railroad at a hearing at Philadelphia the following week. The Speaker of the House of Representatives promised a new legislative Committee on Corporations friendly to the producers' antidiscrimination bill. Campbell planned to subpoena the Pittsburgh Standard Oil men to attend the Titusville hearings. Rejoicing in his success in the role of "peacemaker" between the state administration and the associate counsel Campbell returned to the Oil Region triumphant.[12]

Sherman was undoubtedly elated at this favorable turn of events, since he had penned a pessimistic analysis of the problem confronting the associate Commonwealth counsel. "1. No books—contracts showing rebate. . . . 4. All to be treated alike. Same evidence. 5. Cannot make out case without Exam . . . Combination—Rebate unknown—must be unreasonable."[13]

When the Titusville hearings began Sherman and Shiras represented the Commonwealth, opposing John H. Hampton, representing the Pennsylvania Railroad, D. T. Watson and S. C. T. Dodd, representing the United Pipe Lines, and Charles B. Guthrie, counsel

[12] Sherman Papers, Campbell to Sherman, January 5, 1879.

[13] *Ibid.,* undated pencil MS.

of the Acme Oil Compay. Relying upon Campbell's report of Lear's revised instructions Sherman had subpoenaed John D. Archbold, President of the Acme. He called him to the witness stand. When General Sweitzer read Lear's original instruction as to the procedure to be followed by the Commonwealth counsel, a warm colloquoy precipitated consultation with Lear by telegraph who replied with a substantial confirmation of his original order in telegrams to Sweitzer and to Guthrie, Archbold's counsel. After that, on the advice of their counsel, Archbold and J. J. Vandergrift, President of the United Pipe Lines, refused to answer questions put to them. The defendants were taking full advantage of Lear's departure from the customary procedure in equity proceedings.[14]

At Titusville on January 7 and January 8, however, Sherman and his associates presented crucial evidence for the Commonwealth. Ohlen's Oil City agent testified that the American Transfer Company and the United Pipe Lines had declined his requisitions for tank cars or to load them during the "immediate shipment" crisis. The asserted car shortage had depressed oil prices when production was moderate, only to be succeeded by large railroad shipments when production was at a peak and oil had fallen to 73¢ a barrel. Under cross-examination he admitted that he belonged to the Producers Unions and had contributed to the General Council's Legal Remedies Fund. Mark Acheson's examination and Hampton's cross-examination drew from David Jones, the Reading's Assistant Comptroller, decisive evidence of Pennsylvania Railroad rebating of 64.5¢ per barrel on oil shipments to New York to an undisclosed recipient. Jones's identification of a Mr. Justice, the Pennsylvania Railroad's Assistant Auditor, as cognizant of its rebating opened up a lead that Sherman exploited fully.[15]

After Patterson presented more evidence of railroad discrimina-

[14] *Ibid.*, Sweitzer to Lear, January 7, 1879, Lear to Sweitzer, January 7, 1879, Lear to Guthrie, January 7, 1879, Hartranft to Patterson, January 7, 1879; Commonwealth *v.* Pennsylvania Railroad, *Printed Testimony,* pp. 1–6, 7–12; *Titusville Morning Herald,* January 8, 1879; *History of the General Council,* pp. 22–23.

[15] *History of the General Council,* pp. 23–24; Commonwealth *v.* Pennsylvania Railroad, *Printed Testimony,* pp. 25–31; *Titusville Morning Herald,* January 8–9, 1879.

tions the receiver for H. L. Taylor & Company, Thomas S. McFarland, testified that he had had to sell his Butler and Armstrong county oil at 25¢ less than market price and that he had lost oil from the overflow of his well tanks during the car shortage from May to July 1877. Inability to secure tank cars had obliged him to cancel the sale of oil to the independent Jules Rousseaux of New York and to sell 90% of his product to the Standard Oil. Thus, he had been blocked in his attempt to develop an outside market for the product of this 50% Standard Oil-controlled company.[16] Encouraged by these disclosures Sherman invited the "outside" New York oil men to testify at the hearings,[17] and he demanded copies of the Lake Shore railroad's answer to the Commonwealth's injunction bill.[18]

After adjournment at Titusville Sherman described graphically to Patterson the predicament of the Commonwealth's counsel arising from the conflict between Lear's instructions and the law with the intimation that that fighting member of the producers' triumvirate should do something about it. Compulsory process of witnesses and company records could issue only against Pennsylvania corporations. The defendants' denials in all the injunction suits placed the burden of proof upon the Commonwealth to demonstrate the existence of a combination in restraint of trade to create a monopoly that worked "a *public* injury." Upon that the Suits must stand or fall. Proof of the combination in the form of written contracts was in the possession of the defendants and could not be secured without calling for the evidence they possessed as was customary in such litigation. However, the defendants' witnesses' sworn denials that documentary evidence existed and denials of the validity of signatures on all papers admitted as evidence, and their withholding of

[16] *Titusville Morning Herald,* January 9, 1879.

[17] Sherman Papers, Sherman to Ohlen, January 9, 1879.

[18] *Ibid.,* Sherman to J. W. Welmon and James Mason, January 10, 1879, Sherman to A. M. Martin, February 21, 1879. Sherman retained a certified court stenographer's copy of the testimony taken at the hearings. The published testimony relevant to the bill against the Pennsylvania Railroad Company is incomplete. None of the evidence taken against the other defendants was printed.

essential evidence on these points became a major obstacle to success, since Commonwealth's counsel were not permitted to procure contracts, correspondence, or business records "by the usual equity practice" and were denied the "right to call for any information from the officers, or clerks or agents" of the defendant corporations. As a result the hands of the Commonwealth's counsel were tied and they were "able only to give an occasional kick with one foot at a time." If they subpoenaed the staff of Pennsylvania corporations "not confederates or defendants" who possessed pertinent information, the defendants' counsel claimed the right of cross-examining them on "their private business" in areas "entirely foreign to any issue" in the Suits to the profit of the defendants and injury to those neutral companies! Hence, Sherman concluded sarcastically:

> We have the astonishing spectacle of a suit in Equity progressing before a Master, in which the plaintiff is not allowed to procure any evidence from the defendant upon the ground that its private affairs must not be investigated, while the defendant may pry into and investigate the affairs of other and perhaps rival corporations without limit.[19]

At Sherman's suggestion Shiras presented this situation to Lear, including the refusal of Archbold and Vandergrift to testify because of his instruction and also the defendants' inconsistency in obstructing a full disclosure of the truth after claiming a "*legal right* to discriminate" between shippers. The testimony of Jones, the Reading's Assistant Comptroller, had disclosed that the Pennsylvania Railroad extracted refunds on oil freight-rate receipts from the Reading's proportion of rebates on through shipments. Didn't this fact and the discovery that these rebates were "so large as to crush out shippers" not receiving them justify the Commonwealth "in bringing all the facts to light?" It was proper, Sherman asserted, for the Attorney General to ask the Supreme Court to instruct Archbold and Vandergrift to testify.[20]

[19] Sherman Papers, Sherman to Patterson, January 9, 1879.

[20] *Ibid.,* MS "Points to be made in a letter to be written to Attorney General Lear by Mr. George Shiras," Shiras to Lear, January 10, 1879 (copy).

Patterson took Sherman's letter to him and a copy of the first day's hearing to Hartranft and Lear at Harrisburg. Three days later Sherman received a collect code telegram from "P": "Emerge we meet only Context says Melon occupy must Eulogy record stirred things." Decoded this read: "Prospects are improving. We meet Continental Hotel, Philadelphia Tuesday Says J. D. A. & J. J. V. must testify. Record stirred things."[21] Aside from this, the appeal to Lear was unproductive. Master Sweitzer had no power to commit witnesses for contempt but could only certify refusals to answer questions to the Supreme Court. This made clearer the handicaps under which Sherman and his associates labored. The denial to them of the right of compulsory process and the stipulation that they must make a *prima facie* case against each defendant from the complainant's witnesses before calling upon defendants' officers and records were unprecedented as Sherman asserted. The associate counsel proceeded, despite Ohlen's telegraphed willingness to negotiate "an arrangement with the Standard Oil."[22] Sherman believed, apparently, that despite unprecedented difficulties the Commonwealth could win. Campbell predicated a "voluntary rush" of "all the PRR & Standard people" to "the witness stand to explain & lie away our testimony."[23]

Since Acheson and Shiras lacked detailed knowledge of conditions in the oil industry and of its relations to the railroads, Sherman was called upon during the examination of witnesses to develop the more difficult phases of the presentation of complainant's evidence.

Patterson brought to him lengthy instructions from Lear to Sweitzer and the associate Commonwealth counsel that approved what Sherman had done previously. Archbold and Vandergrift agreed to testify. Sherman questioned the Allegheny Valley Railroad officials with excellent effect. However, accumulated obligations of his practice prevented him from questioning witnesses at the hearings again until February 15 at Philadelphia, but he contin-

[21] *Ibid.,* Patterson to Sherman, January 10, 13, 1879.

[22] *Ibid.,* Ohlen to Campbell, January 14, 1879.

[23] *Ibid.,* MS note [Campbell to Sherman]; *History of the General Council,* pp. 22–24.

ued to direct the presentation of the Commonwealth's cases against the four defendant railroads and the United Pipe Lines.[24]

When pressure of other obligations prevented Sherman from attending the first hearings at Philadelphia Campbell wrote him that "the examination goes on badly without you." Sherman was needed to question Archbold and Vanderbilt, for which Shiras was inadequate, although the latter got on well enough with the New Yorkers, Ohlen and Lombard. John Scott, Campbell added, knew of Cassatt's "interest in the Lubricating Oil Works in Oil City, Can you have the matter out. Tell Patterson to be sure to be [at] Phil." This note impugned the motives of railroad officials who granted large rebates to the Standard Oil. With Campbell's and Patterson's assistance Acheson and Shiras were not ineffective in Sherman's absence. From the Auditor of the Allegheny Valley Railroad they extracted proof that it paid a 15¢ a barrel rebate to the American Transfer Company. "Things is working," Campbell informed Sherman after this. John Scott, President of the Allegheny Valley Railroad, promised more evidence when it was desired.[25] After the hearings were adjourned to Pittsburgh for presentation of the Commonwealth's case against the United Pipe Lines, an explosive letter from Patterson on February 9 informed Sherman that his fellow triumvirs did not team well with his associate counsel. Campbell was "in a hurry to *close* the case" and had listened "to Mat Quay and that crowd about compromises &c &c." He did "many indiscreet things," Patterson added, "and has got the case a - s foremost—Shiras stopped at request of Govr at Harrisburg, and consequently his conservatism is increased. He is very timid about closely examining Standard men because they are *our* witnesses & will *not* push them fearing they will lie, and that we may not contradict them."

Patterson had to appear before the Senate Committee in Washington in behalf of the interstate commerce bill, but would be in

[24] Sherman Papers, Campbell to Sherman, January 18 (twice), 1879, Shiras to Sherman, January 18, 1879, Patterson to Sherman, January 18, 1879, Sherman to Campbell, January 28, 1879, Sherman to Sweitzer, January 27, 1879, Sweitzer to Sherman, January 21, 1879.

[25] *Ibid.,* Campbell to Sherman, January 26, n.d.

Pittsburgh immediately afterward. He maintained that "this case *must be* proven so far as the conspiracy is concerned by calling ringsters, and they have finally agreed to start now where I did when we issued the first subpoena—with the S. I. contract of 1872." Acheson or Sherman should summon Archbold and Vandergrift again to the witness stand, and Sherman and Campbell should question the latter "on the Pipe Line Case on Thursday morning." Unless Sherman appeared, "there will probably be nothing done." [26]

At the Pittsburgh hearings the manager of the Fox Farm Pipe Line demonstrated the idleness of many Allegheny Valley tank cars in June 1878 when W. J. Brundage, manager of the Green Line (partly owned by the Standard Oil through the Car Trust) on the Pennsylvania Railroad was filling only small portions of car orders for shipments to the "outside" New York buyers and refiners. Samuel Stowell, proprietor of Stowell's *Petroleum Reporter* and oil expert of the United States Treasury Department's Bureau of Statistics disclosed the swift decline in oil refining in the Oil Regions from 1877 to 1878 (from 13.2% of the total oil shipped to 2.2%), while New York crude-oil receipts rose from 35.2% to 42.9% of total shipments, and Pittsburgh's rose from 14.1% to 17.1% while Cleveland's fell a sixth. New York City had emerged the victor in the metropolitan contest for supremacy in the oil trade as a by-product of the Standard Oil's triumph in the oil rate war of 1877. This was profoundly disturbing to the Oil Region and Pennsylvania citizens elsewhere.[27]

At the first Philadelphia hearings Shiras had extracted from Lombard invaluable data on the relation of Erie Ring-rate favoritism to the origin of the Standard Oil. Lombard supplied historical data validating the assertion that the Standard Oil's primacy as a crude-oil shipper derived from rebates, to which trunkline pools contributed. On crude-oil shipments to New York the Standard

[26] *Ibid.*, Patterson to Sherman, February 9, 1879.

[27] *Ibid.*, Sherman to A. M. Martin, February 21, 1879; Commonwealth *v.* Pennsylvania Railroad, *Printed Testimony*, pp. 65–77, 124–128.

Oil's success since 1872 had depended upon rebates, a thesis that Nevins validates.[28] Lombard and George F. Gregory, independent Brooklyn refiner, disclosed that the Standard Oil exclusive ownership and use of tank cars on the New York Central and Erie railroads and its lease of the latter's Weehawken oil docks had made the New York independents dependent upon the Pennsylvania Railroad. Subsequently they and Ohlen described the refusal in 1878 of Thomas A. Scott and Cassatt to give them more tank cars from those standing idle in their yards because the Pennsylvania Railroad lacked stomach for another battle with the Standard Oil, and because the former recommended that they make "an arrangement" with it. That suggestion Ohlen had rejected.[29]

Cassatt had declined to give an "outside" Brooklyn refiner the same low rate on crude-oil shipments as the Standard Oil, even if the former shipped as much oil because the latter performed the indispensable function of harmonizing the trunklines of the oil-traffic pool. R. T. Bush, another Brooklyn independent refiner, disclosed that Cassatt had threatened to purchase the Equitable pipeline if the New York independents became identified with it and to reduce any other independent pipeline to "old iron." Yet, since the spring of 1877 the New York independents had been the largest shippers of crude oil to New York City on his railroad. Most of this evidence was corroborated subsequently by testimony from Cassatt, Archbold, and Vandergrift.[30]

Despite Campbell's and Patterson's telegraphed pleas Sherman remained in Titusville until mid-February. Among his other professional obligations was counseling the Titusville and other local Producers' Associations against contracting with the American Petroleum Company as their proposed exclusive selling agency, viewing such action as wildly imprudent, certain to end in disaster, and

[28] *Study in Power,* I, 149, 187–188; Commonwealth *v.* Pennsylvania Railroad, *Printed Testimony,* pp. 253–275.

[29] Commonwealth *v.* Pennsylvania Railroad, *Printed Testimony,* pp. 161–162, 166–168, 182–183, 197, 200, 253–257.

[30] *Ibid.,* pp. 160–161, 205–206, 237–241, 269–271.

bound to prevent producers from engaging in refining or from aiding independent pipelines.[31]

Sherman had been disgusted on January 15 by an acrimonious debate in the General Council of a proposal to initiate negotiations with Rockefeller, which that magnate had invited as the Suits began to uncover sensational data on railroad discriminations favoring his organization. Simultaneously, the *Titusville Morning Herald* criticized the producers for inadequate cooperation with the General Council's curtailment of drilling, attacked its appeal to Congress for an interstate commerce law, and defended the railroads' rate favoritism to the Standard Oil as justified by its larger shipments. This undercut two of the Council's policies and intimated that the Republican party was weakening in its opposition to railroad discriminations.[32]

Privately at this time Sherman lost confidence in the oil business. He informed C. B. Seymour, his manager, that he wanted to get out of it at the earliest moment.[33] His misgivings were produced by Campbell's vacillation in the conduct of the Commonwealth Suits and by the producers' short-sighted profit drive as expressed in the General Council which threatened to defeat their extraordinary program. On February 18 the *Morning Herald* published a German oil dealer's skepticism of the Producers Unions' prospects of success in their struggle with the Standard Oil and the railroads. He doubted their ability to curtail production, to hold together, and to compel the carriers to give equal treatment to shippers.

At Oil City where Vandergrift was powerful, the local Producers Union appointed a "compromise committee" which met with a committee of managers of the Producers Associations that were attempting to retire the oil surplus. At that early March meeting John Fertig, E. O. Emerson, and J. L. McKinney of Titusville, Sherman's friends, led in the discussion of the oil production problem that reviewed a recent "now famous oil conference" in New

[31] Sherman Papers, Campbell to Sherman, February 4, 6, 1879, Patterson to Sherman, February 12, 1879, Sherman to T. C. Joy, January 27, 1879.

[32] *Titusville Morning Herald,* January 12, 16, 28, 1879.

[33] Sherman Papers, Sherman to Seymour, February 3, 1879.

York where "a few leading producers and the refiners" had agreed that the refiners and producers jointly should make the best terms possible with the transportation companies if "cooperation were . . . determined upon" and rates should be "free, equal, fair and open to both sides, while the pending suits against the railroads would not be interfered with in any way." However, "cooperation" was Rockefeller's favorite term for his method of dealing with crude-oil marketing and railroad rate problems. The tentative scheme may have been a preliminary to negotiations for out-of-court settlement of the Suits and undoubtedly contributed to Sherman's discouragement.[34]

However, on March 12 the General Council rejected that conference's proposal by a vote of 63 to 3 and upheld Sherman's leadership in the Suits while expressing the independent producers' profound distrust of the Rockefeller coterie. Campbell reported jubilantly that Governor Hoyt backed the producers whose case against the railroads was now fully developed before Master Sweitzer. The courts "will strike the compact" of the Standard Oil with them "like lightning from heaven, the Standard will writhe among her worshippers, and the Producers will rise disenthralled, and the masters of the situation," he predicted. Admitting that the carriers' influence at Harrisburg made it another Sodom or Gomorrah and the Committee on Legislation's antidiscrimination bill hadn't "a ghost of a chance" there, he predicted that relief "will come through the courts."[35]

In renewed attempts to negotiate a cooperative settlement the Standard Oil agreed grudgingly to permit the independent transportation routes to seaboard via the Equitable pipeline and the new but incomplete Tide Water. Benson's refusal to join[36] defeated this attempt to achieve what in 1872 had been the declared, comprehensive objective of the South Improvement Company, the inclusion of all branches of the oil business in a single organization. Roger Sherman was the Tide Water's general counsel, and to him must be

[34] *Titusville Morning Herald,* March 5, 6, 1879.
[35] *Ibid.,* March 12, 1879.
[36] *Ibid.,* March 22, 31, 1879.

attributed some responsibility for Benson's rejection of what was undoubtedly a deft Standard Oil bid to take over control of the entire industry, since in any "cooperative" plan its preponderant position would necessarily receive due weight. The project's proposal and its serious consideration by leading producers and the New York independents necessarily weakened the position of the Producers Unions and jeopardized the Commonwealth Suits obliquely. It had pitted the managers of the Producers Associations against the General Council and contained implications which the Cameron Republican organization was certain to note.

The General Council's legislative program fared badly at Harrisburg thereafter, except for successful opposition to a proposed state tax on oil production. To defeat it, enormous energy had to be rallied among producers. That diverted support from Lewis Emery, Jr.'s antidiscrimination bill that now included pipelines. It was defeated although he fought hard for it with *Morning Herald* support. Thus, long before May, it was obvious that the independent producers were meeting growing opposition within the Republican party, which strengthened Secretary of State Matthew S. Quay's pressure upon them to negotiate a settlement of the Commonwealth Suits.

Although Sweitzer's hearings continued at Pittsburgh after mid-February and Campbell desired his attendance, Sherman was obliged to attend court instead at Meadville. Then, from February 22 to February 24 Campbell was put on the witness stand and questioned searchingly by Acheson amid Hampton's extraordinary objections. Campbell described the character of the Producers Unions, their General Council, how he and Patterson had temporarily persuaded the Pennsylvania Railroad to secede from the trunkline oil-traffic pool in 1877, and the origin of the Commonwealth Suits. Campbell's importance as a witness derived from his fighting leadership of the General Council and his large-scale producing operations. For years he had been the Empire Transportation Company's "heaviest customer." His testimony provided an exciting climax to the proceedings and was highly embarrassing to the Pennsylvania Railroad and the Standard Oil.

*Ex parte* but well maintained during severe cross-examination this testimony divulged the Standard Oil's control of the Pennsylvania Railroad's tank cars and the crude-oil market and the squeezing process whereby the railroads had obliged independent pipelines to sell out to the United Pipe Lines. Receipt of exorbitant, decisive rebates gave the Standard Oil its control over the market, Campbell asserted, thus supporting the Commonwealth's central thesis in the Suits. The Standard Oil enjoyed similar rebates in every other state, he added. He denied that the railroads had a legal right to so discriminate between shippers as to give any party "the monopoly and control of our trade." All shippers of carload lots should be given the same "wholesale" rates. He asserted vigorously the long- and short-haul principle that would bulk so importantly in subsequent railroad regulation. He denounced the Pennsylvania Railroad's boycott of the Equitable-Erie Canal route and asserted that it "is today a pensioner at the hands of the Standard Oil Company."

Overproduction would not exist, he claimed, if the market were "free," although the producer would have to accept a lower price. A "competition of buyers" and opportunity for producers to erect their own refineries if dissatisfied with crude-oil prices were what he sought. He disclosed the tremendous spread between the price of crude oil and of refined oil in 1876 ($4.20 for crude in the Oil Regions versus $16 a barrel for refined in New York City) as the product of Rockefeller's near monopoly of that year. Campbell described the virtual extinction of oil refining in the Oil Region which the Standard Oil victory of 1877 had wrought. He discussed its large producing activity "in the northern district."

After divulging the sources of financial support of the Commonwealth Suits, he stated that they were not directed primarily against the Pennsylvania Railroad. He urged that victory in them would revive its control of the oil traffic by stimulating the erection of new independent refineries in west Pennsylvania. However, his admission that the General Council was very short of funds played directly into the hands of the defense. Stated parenthetically during Hampton's cross-examination, this encouraged the defendants to revive the strategy of delay as a means of forcing a favorable out-of-court settlement of the Suits.

Campbell's stout assertion that the Standard Oil lacked any technological advantage over the average, well-financed refiner while it was handicapped by the obligation to pay large sums monthly to refiners whom it had forced out of business gave strong support to the proponents of the competitive system. Certainly his long testimony provided ammunition for the critics of the Standard Oil for decades to come. Especially effective was his demonstration that the trunkline-Standard Oil pool had deprived Pittsburgh and Philadelphia of their natural advantages for the refining and export of petroleum and its products, while giving an artificial primacy to New York City.[37]

With Campbell's testimony Sherman and his associate counsel completed their *prima facie* case against the Pennsylvania Railroad and secured much evidence applicable to the United Pipe Lines Company. They then subpoenaed Cassatt, John Scott, and certain Pennsylvania Railroad subordinates and records in pursuance of Lear's revised instructions which continued to guide the prosecution under Governor Hoyt.

Sherman and Shiras put J. J. Vandergrift, President of the United Pipe Lines Company, and William Frew, a director of the Standard Oil Company of Pennsylvania, on the witness stand. They described the Standard Oil's selling agency for the refiners' pool, Rockefeller's and Flagler's negotiations for freight-rate concessions, the monopolistic refiners' agreement in Pennsylvania, and the Standard Oil of Pennsylvania's acquisition of the independent refiners of the Oil Region. Vandergrift and Frew perjured themselves at times. Sherman certified Frew to the Supreme Court for refusing to answer important questions. Both Standard Oil executives invoked technicalities to avoid clarifying the internal character of the Standard Oil combination. Regarding the Acme Oil Company, a Standard Oil subsidiary, Shiras elicited from John D. Archbold significant admis-

[37] Sherman Papers, Patterson to Sherman, February 8, 12, 19, 20, 1879, Sherman to Shiras, February 20, 1879; Commonwealth *v.* Pennsylvania Railroad, *Printed Testimony,* pp. 579, 581–588, 295–306, 313–315, 558–615.

sions of the receipt of rebates and data on the oil combination's marketing organization in New York City. The Erie Railroad's monthly statements to the Acme of rebates paid on oil shipments to Communipaw enlarged the record on rebates significantly.[38]

Spectacular evidence of rebating's decisive advantages for the Standard Oil was brought to light in the examination of the officials and records of the Pennsylvania Railroad. Most sensational was the discovery that the American Transfer Company collected 22.5¢ per barrel "commission" from the Allegheny Valley Railroad *on all oil* carried by it as compensation for allowing it a pro rata share of this traffic to Pittsburgh, which was obviously based upon South Improvement Company precedent. Then Robert W. Downing, the Pennsylvania Railroad's Comptroller disclosed evidence of frequent rebating when he testified that he paid *"about 40,000 rebate vouchers a year of all kinds."* This preceded the Hepburn Committee's discovery that the New York Central had 6,000 rebate contracts in force. Jefferson Justice, the Pennsylvania Railroad's Assistant Auditor, adduced documentary evidence of large and varying rebates paid to favored shippers.[39]

A. J. Cassatt's testimony provided invaluable "inside" evidence on the Pennsylvania Railroad's past policy toward the oil traffic. He described the payment of large rebates to secure oil traffic during the rate war of 1877, the total defeat in that contest, and the 22.5¢ a barrel "commission" on *all oil shipments* extorted by the American Transfer Company afterward. He admitted that after further rate concessions the Standard Oil paid in October and November 1878 only 38¢ a barrel on crude-oil shipments to New York harbor, as the result of rebates and commissions paid it, instead of the open rate of $1.15. That 38¢ the Pennsylvania had to pro rate with the Lehigh Railroad which carried the oil for the latter part of the shipment. There were understandably, Cassatt admitted, no "outside shippers of refined oil" on the Pennsylvania.

Of great importance for the Commonwealth's case was Cassatt's admission that the four eastern trunklines determined oil freight

[38] Commonwealth *v.* Pennsylvania, *Printed Testimony,* pp. 369–387, 416–435, 532–535.

[39] *Ibid.,* pp. 500–509, 629–658, 737.

rates jointly and that this pool was warring upon the Equitable's route to the seaboard. All this disclosed a conspiracy to create and perpetuate the Standard Oil's monopoly. Cassatt admitted that the Pennsylvania had taken the initiative in reviving and perpetuating the trunkline oil-traffic pool from 1877 to 1879. He denied, however, that any Pennsylvania official was "interested in the Standard Oil Company." [40]

However, when examined by Wayne MacVeagh, a leading Philadelphia attorney acting as an additional counsel of the Pennsylvania Railroad, Cassatt declared that that carrier desired equal open rates for all shippers if its competitors were required to charge the same. Under such conditions "we can always secure our fair share of the business," he said. He defended the trunkline pool as a preventative for competitive rate-cutting but admitted that pooling agreements were difficult to enforce. For that purpose he desired a national statute.[41]

This testimony virtually completed the Commonwealth's case against the Pennsylvania Railroad. It advanced the cases against the other defendants materially. The Commonwealth associate counsel were encouraged by Cassatt's concluding statement, since a Supreme Court ruling obliging the trunklines to treat all shippers equally was a major objective of the Suits. As for the Standard Oil, chief beneficiary of the alleged conspiracy, conclusive evidence had been adduced that described John D. Rockefeller's and Henry M. Flagler's personal negotiation of railroad rate favors, while many features of that monopolistic combination had been divulged.

Campbell concluded that the Commonwealth now possessed sufficient evidence to institute criminal proceedings against the Rockefeller brothers, Flagler, Charles Pratt, Charles Lockhart, Frew's partner and a shareholder in the Standard Oil of Pittsburgh, Bostwick, Warden, Vandergrift, and O'Day. Campbell and Shiras recommended that an information be filed in Clarion County Quarter

[40] *Ibid.*, pp. 614–616, 661–726, 732–756; *History of the General Council*, p. 24; *Argument of Mr. Franklin B. Gowen*, p. 7.

[41] Commonwealth *v.* Pennsylvania Railroad, *Printed Testimony*, pp. 729–730.

Sessions charging the Standard Oil with conspiracy to monopolize the oil trade, to defraud and oppress the producers, to deprive the carriers of their just earnings, and to crush all competing refineries. Campbell and Shiras urged that Sherman utilize George A. Jenks's ability as a criminal lawyer and institute the criminal proceedings immediately. Such action, they predicted, would separate the railroads from the Standard Oil and defeat the attempts of the Oil City committee to negotiate a compromise with Rockefeller. A criminal conspiracy trial of the Standard Oil executives, Shiras predicted in agreement with Sherman's original view, would "produce speedy results." Shiras and Acheson then agreed to present the evidence in the United Pipe Lines, Atlantic and Great Western, and Dunkirk railroad cases before Master Sweitzer at Titusville.[42]

Campbell and Shiras wrote to Sherman in code, who agreed with these recommendations in a judicious, good-humored reply. He expected to complete presentation of evidence in the other injunction cases and the *quo warranto* against the United Pipe Lines in a short while. He advised filing an injunction bill against the American Transfer Company since it appeared to be a kind of "Credit Mobilier for the Combination," and connected the United Pipe Lines' system with the railroads. Similar proceedings should be instituted against the Standard Oil of Ohio and the New York, Lake Erie & Western Railroad (the old Erie Railroad), taking care in each suit to include "every branch, by whatever name known, of the Combination." Since neither of these was a Pennsylvania corporation he believed that Governor Hoyt and his Attorney General would consent. He promised to direct Jenks to be ready "for action," and asked Shiras to compose the developing differences between Campbell and Patterson. For skillful candor this astute letter to Shiras had few equals. Sherman advised study of the evidence gathered against the Pennsylvania Railroad for guidance in presentation of testimony in the other cases before General Sweitzer. In each proceeding the Commonwealth should drive directly to the point.[43]

[42] Sherman Papers, Campbell and Shiras to Sherman, March 16, 1879.
[43] *Ibid.*, Sherman to Shiras, March 17, 1879.

# CHAPTER VII

# *To the Verge of Victory*

WHILE in charge of the Commonwealth Suits Sherman evinced great respect for Shiras's ability and appreciated how his influence with the Governor could be employed to expand the scope of the Suits. In March 1879 Shiras negotiated for Sherman with Henry W. Palmer, the new Attorney General, for permission to institute a *quo warranto* proceeding against the American Transfer Company. Palmer consented conditionally, if it could be made "reasonably clear . . . that the Commonwealth can win." [1]

Campbell and Patterson cooperated with Sherman in developing the Suits, but not without friction with each other. Patterson telegraphed Sherman in mid-March that he would oppose "Campbell's proposition" of a criminal proceeding at a projected Pittsburgh meeting of the producers' leaders with the other associate counsel. Instead, Patterson suggested that O'Day, Campbell, Vandergrift "& any others that you want," be subpoenaed for a Titusville hearing before General Sweitzer. Campbell also advised Sherman to subpoena the officials of corporations in the oil combination.[2] Campbell and Vandergrift were recalled to testify with reference to each Suit.

General Sweitzer's irritation at the defense counsel's delaying tactics enabled Sherman to cultivate a sympathetic *rapport* with him. After consenting to one unnecessary delay the Master wrote to notify him of the changed schedule and added: "Now. Get a service on the cusses, and let's bring them up." The impatient Campbell wrote Sherman, "We must overthrow these thieves next week or

[1] Sherman Papers, Palmer to Shiras, March 1879.

[2] *Ibid.,* Campbell to Sherman, March 25, 1879, Patterson to Sherman, March 18, 19, 1879.

never."[3] The *Titusville Morning Herald* had just published the letter from Daniel O'Day to Cassatt of February 15, 1878, demanding a commission of 20¢ a barrel on all oil shipments on the Pennsylvania Railroad for the American Transfer Company as compensation for "protecting" it "in a paying rate of freight for the oil it carries."[4] E. C. Hoag of the Pennsylvania Transportation Company advised Sherman that the American Transfer Company was owned by a few of the "innermost ring of the S. O. Co.," and that the connection of its pipes leading from the United Pipe Lines Company's gathering system with the railroads should be brought out in the hearings. This explained O'Day's success in wringing the commission from the Pennsylvania Railroad.[5]

A scheduled sensation was prevented when the Atlantic and Great Western Railroad officials refused to testify or produce subpoenaed documents. For this refusal they were certified to Justice Trunkey of the Pennsylvania Supreme Court. The refusal of certain shippers to testify against that railroad, which claimed to be beyond the court's jurisdiction, was also certified.[6]

During the April 10 and April 11 hearings the testimony of James B. Campbell, a Titusville producer, established despite vigorous cross-examination that the United Pipe Lines' "immediate shipment" policy resulted from a "plan adopted to force producers to sell at much lower figures than the open market warranted." Vandergrift and O'Day had refused to answer questions on the subject.[7]

The *Morning Herald* published an inimitable description of those Master's hearings. It declared that the prisoners at the bar were "great corporations . . . through whose lines wealth runs like Pactolus over golden sands." The article continued:

Here is the majesty of the Supreme Court of the great Commonwealth, acting by its deputy. There sits the Master, General Sweitzer, elderly

[3] *Ibid.,* Sweitzer to Sherman, March 26, 1879, Campbell to Sherman, March 27, 1879.

[4] March 28, 1879.

[5] Sherman Papers, E. C. Hoag to Sherman, March 28, 1879.

[6] *Ibid.,* MS memorandum in Sherman's hand, Campbell to Sherman, March 31, 1879, Sherman to John Trunkey, April 11, 1879.

[7] *Titusville Morning Herald,* April 12, 1879.

but vigorous, bald as all become with hard thinking, with a mild blue eye, and courteous manner. The chief counsellors—Dodd and Sherman—sit opposite each other, and opposite they are. The witness is at the table. The stenographer—Mr. Martin—is there. The short-hand writer of the Titusville HERALD is also there. The spectators lay around like sinners at a camp-meeting. Some draw near as if the mystery was all coming out with the next breath and sentence; others sit in remote seats and simply come to gaze and dream.

Sherman's manner seems to say to witness, "I want your secret." Dodd's, "Get it if you can." There is langour in the air. There is no hurry. . . . O, how densely ignorant these witnesses, how mildly innocent, how very careful! . . . This is the way with all Courts, reminding one of "the lazy Schield or wandering Po" in Goldsmith's Traveler.

The counsel for the Commonwealth, looking up, asks a question. . . . The witness looking around to *his* counsel, *his* counsel also looks up, as if to penetrate a mystery. A pause. The counsel for the defendant seems to be weighing the question in his mental scales. The Master seems to be analyzing it. Counsel for the defendant at last nods. Master nods. Witness thus encouraged prepares to answer. Counsel for Commonwealth bends backward. Counsel for defendant leans backward. Spectators in back row wake up. Witness begins. "Hold on!" cries the alarmed counsel for defendant. "Go on!" says the delighted counsel for the prosecution. The witness looks confused. The Master is lost in thought. The spectators get excited. And then follows argument, reference to papers, accounts, affidavits, session laws, authorities. Counsel for defendant gets sarcastic; counsel for Commonwealth laughs. Everybody laughs. The Master tells a story. Stenographer yawns, spectators go to sleep.

Suddenly all wake up to business again, the thing is getting hot; dark references to occult documents are made . . . these men contend like Heenan and Sayers, Macbeth and Macduff, fighting for a crown! A great deal may depend on what is brought out.

You shall see what corporations can do, and what States can do, whether corporations are fiery dragons wound round and round the majestic sinews of the State, or like only supporters of the loins, breastplates and necklaces, to be worn or laid aside at the pleasure of the sovereign people.[8]

[8] April 11, 1879.

Sherman had difficulty in completing the Commonwealth's case against the United Pipe Lines. With Campbell's help he subpoenaed witnesses cognizant of the pipeline company's management practices and "immediate shipment" policy.[9] Shiras and Acheson believed that a Supreme Court Justice in chambers had "all the powers of the court, touching rules or orders and other proceedings."

Justice Trunkey obliged them by ruling favorably on the issues certified to him by General Sweitzer. Although Trunkey ruled that the Master possessed no compulsory process to enforce attendance and that witnesses could decline to answer questions about immaterial or irrelevant matters, they should answer "*prima facie* proper" questions. The Commonwealth could go back several years to prove acts done in furtherance of the conspiracy. Each specific act in furtherance of the conspiracy "need not be charged in the bill," although it must be amended before questions could be asked regarding the United Pipe Lines' overissue of oil certificates and "*lending* of oil." Trunkey also ruled that the Commonwealth could demonstrate the connection between the defendants and the conspiracy by proving stock ownership in the corporations in the combination. Trunkey granted a rule requiring that the Atlantic and Great Western Railroad show cause why an order should not issue upon its Auditor and Treasurer "to produce vouchers &c & answer questions." The Supreme Court would consider its reply on May 8.

Sherman promised to go over the Pennsylvania Railroad case carefully to "see whether the testimony cannot be closed on our part." He would meet Acheson and Shiras at Pittsburgh "*early next week for consultation.* With proper management, we can soon see the beginning of the end," he observed to Shiras.[10] Acheson and Shiras agreed that it was important to close testimony in the case against the Pennsylvania Railroad as soon as possible.[11]

Campbell was the Danton of the producers' movement. He

[9] Sherman Papers, Sherman to Campbell, April 15, 1879, Sherman to E. Crossman, April 15, 1879, Sherman to W. R. Weaver, April 15, 1879.

[10] *Ibid.,* Sherman to Shiras, April 18, 1879.

[11] *Ibid.,* Shiras and Acheson to Sherman, April 14, 1879.

was bitterly opposed to further postponement of the hearings and protested vigorously a ten-day postponement because of S. C. T. Dodd's "sick child." That meant "ruin to us," he wrote Sherman indignantly, and "safety to them, the members of the oil combination." This was due to the shortage of funds of the Petroleum Producers Unions and the swelling glut of Bradford oil from whose effects victory in the Suits was expected to bring material relief. Campbell declared that Trunkey's ruling that Master Sweitzer lacked compulsory process "had shorn the Suits of their strength" and handicapped the producers in their "warfare with the most evil opponent any litigant ever met." He urged Sherman that "like men" they should "strike a blow which our enemies . . . will feel to their very marrow,"—have them arrested on criminal charges which the entire movement demanded. This the Associate Commonwealth Counsel and the Committee on Legal Remedies must undertake unless they "wish to stand alone without any support." *Toujours l'audace!* Since minor employees of the oil combination had been jailed for attempting to block the Tide Water's pipeline, "we surely on better proof, and with a better case can dare to seize the head devils," Campbell asserted confidently.

He was impoverished after devoting a year and a half to the producers' cause. "The only reward any of us can receive will be in the consciousness that our efforts have freed our courts from this gang of thieves," he remarked. Knowing how carefully Sherman prepared for his litigation Campbell asked him "in this instance to take some chances justified by the situation & trust somewhat to the goodness of our cause & the existence of God who overrules." He demanded closing of the case against the Pennsylvania and "a criminal information against the Standard Directors & Dan O Day for conspiracy." Would Sherman meet with his associates and telegraph him their decision at Parnassus? Campbell reminded him that his legal competence and devotion "to our Cause have been & ever will be my chief reliance." [12]

Trunkey's rule regarding the officers of the Atlantic and Great

[12] *Ibid.,* Campbell to Sherman, "April 1879."

Western Railroad produced an unexpectedly favorable result. That carrier's receiver, John H. Devereux, informed Sherman that he would not oppose a uniform decree against the carriers. This suggested the possibility of a negotiated settlement between the Commonwealth and the defendant railroads. This Sherman and Shiras capitalized upon.[13]

At Pittsburgh on April 22 Sherman consulted Shiras, Acheson, Jenks, and the General Council's Committee on Legal Remedies. His long memorandum to them reported the status of the Suits, recommended additional proceedings, and stated the policy to be followed. The case against the Pennsylvania Railroad was completed. That against the United Pipe Lines could be completed shortly. The acceptance by the Supreme Court of jurisdiction over the Atlantic and Great Western Railroad in receivership was necessary to developing that case. Testimony in the Lake Shore case, he said, "has not been taken" but much taken in other cases would apply to it. Getting evidence directly from its officers would be difficult because of their out-of-state residence. Upon doing so depended closing the case against the Lake Shore subsidiary, the Dunkirk, Allegheny & Pittsburgh Railroad Company.

Acting under Campbell's orders Sherman carried the conference for prosecuting the chief executives of the Standard Oil for criminal conspiracy, and for instituting *quo warranto* proceedings against the New York, Lake Erie & Western Railroad Company, the Buffalo, Bradford and Pittsburgh Railroad Company, and the American Transfer Company. An injunction proceeding against the latter was also agreed upon. These actions, he felt, should be instituted if possible and "a definite conclusion reached in May before the beginning of the shipping season." [14]

The strategy that Sherman recommended was to split the trunkline railroads and their subsidiaries from the Standard Oil and its pipeline subsidiaries by means of the criminal prosecution of the oil

[13] *Ibid.*, S. P. McCalmont to Sherman, April 21, 1879, Sherman to McCalmont, April 21, 1879.

[14] *Ibid.*, MS Memorandum, April 21, 1879, Sherman to Campbell, April 21, 1879; *Titusville Morning Herald,* May 7, 1879.

combination's chief executives. By concentrating the attack upon the Rockefeller coterie and avoiding all especial hostility against the railroads this could be accomplished, since Cassatt and other trunkline officials were willing to abandon "the rebate system." This was practicable since the parties to the trunkline-Standard Oil combination were members of a criminal conspiracy indictable at common law. In order to divide the defendants Sherman held that the railroad company executives should be omitted from the criminal prosecution. The Committee on Legal Remedies and representatives of the General Council except Patterson agreed with Sherman's strategy and approved the criminal prosecution of the Standard Oil executives. Patterson then withdrew from all further connection with the Commonwealth Suits.[15]

Clarion County was regarded as appropriate for prosecution of the criminal conspiracy case because "more overt acts of the alleged conspirators" had been committed there than elsewhere. The Presiding Judge of Quarter Sessions of Clarion County was George A. Jenks's brother! Campbell and Jenks filed the criminal information, swore out the warrants, and informed Sherman happily: "The deed is done, for weal or woe." Campbell regretted Patterson's withdrawal.[16] Sherman informed Governor Hoyt and Attorney General Palmer of the tactical purpose of the criminal conspiracy case—to enable the trunkline railroads to throw off the oil combination's "yoke" and abandon the rebate system. The producers generally wished to place Standard Oil executives behind prison bars as Allan Nevins asserts.[17]

On April 29 the Clarion Grand jury indicted the executives of the Standard Oil, United Pipe Lines, and American Transfer Company on eight counts for conspiracy to create a monopoly of the oil business to the injury of their competitors and of the producers and of the carrying trade of the Allegheny Valley Railroad and Pennsyl-

[15] *History of the General Council,* p. 33.

[16] *Ibid.,* pp. 33–34; Sherman Papers, Campbell to Sherman, "April 1879," Sherman to G. A. Jenks, April 25, 1879.

[17] *Study in Power,* I, 313; Sherman Papers, MS Memorandum, May 9, 1879.

vania Railroad companies, by the extortion of rebates and "by fraudulent means and devices to control the market prices of crude and refined petroleum and acquire unlawful gains thereby." John D. and William Rockefeller, Jabez Bostwick, Daniel O'Day, William G. Warden, Lockhart, Flagler, Vandergrift, Charles Pratt, and George Girty, cashier of the American Transfer Company were those indicted. Their conspiracy's origin was located at Philadelphia in the summer and autumn of 1877 followed by overt acts in Clarion County and elsewhere. The Rockefellers, Bostwick, Flagler, Pratt, and Girty were nonresidents, but all except Girty possibly had been in Pennsylvania repeatedly in advancing the objectives of the conspiracy.[18] O'Day, Lockhart, Vandergrift, and Warden appeared and gave bail. The trial of the case was scheduled for the August sessions.[19]

On May 3 Shiras and Acheson closed the case against the Pennsylvania Railroad Company.[20] Sherman supervised and assisted Jenks's preparations for trial of Commonwealth *v.* Rockefeller *et al.* with certified copies of all the Standard Oil papers filed with the Ohio Secretary of State and all testimony taken in the Commonwealth Suits to date. He urged that the nonresident defendants be extradicted. He prepared to challenge the legality of the refusal of the receiver of the Atlantic and Great Western Railroad to answer questions because the Supreme Court had no jurisdiction over it.[21] The receivership had been instituted by the Philadelphia Court of Common Pleas.

On May 9 Sherman informed Governor Hoyt and Attorney General Palmer of the Associate Commonwealth Counsel's strategy in the Suits. Hoyt agreed that the disposition of the defendant railroads was most hopeful. He ruled that a simultaneous uniform consent injunction decree should be secured if possible as "the only

[18] Sherman Papers, Sherman to Ohlen, May 5, 1879, Campbell to Sherman, April 25, 1879; *History of the General Council,* pp. 33–34.

[19] *History of the General Council,* p. 34.

[20] Sherman Papers, Sherman to Campbell, May 1, 1879, Sherman to Shiras, May 1, 3, 1879, Shiras to Sherman, May 3, 5, 1879.

[21] *Ibid.,* Sherman to Secretary of State, Ohio, May 5, 12, 1879, Sherman to Jenks, May 5, 1879, Sherman to Gilbert, May 6, 1879.

practicable and effective solution of the difficulty." Sherman secured from Palmer *carte blanche* to conduct the Suits "within reasonable bounds." [22] The Supreme Court took under advisement Sherman's application that it compel the Atlantic and Great Western Railroad witnesses to testify. He returned to Titusville to report to the Committee on Legal Remedies.[23]

He then attempted to develop the testimony against the United Pipe Lines to a point that would enable him to close the Commonwealth's case against it also. He presented four producers to Master Sweitzer who testified to the United Pipe Lines' abuse of "immediate shipment" and its arbitrary refusals to run the oil that it had purchased. C. B. Seymour, Sherman's father-in-law, was the most explicit in describing these abuses and the United Pipe Lines' preference for Standard Oil-owned oil in operating its lines. By such means Sherman disclosed the inconsistency of the United Pipe Lines' administration, its abuses, injustices, robbery of the helpless producer, and exploitation of his tanks for storage without paying him storage rates. Sherman then received a request from Palmer that he defer taking testimony from O'Day and Vandergrift until after August court at Clarion County since they were indicted there. Palmer had heard bitter Rockefeller complaints that the civil Suits were being used to secure testimony from the indicted men "to their prejudice." [24]

Because of Rockefeller and railroad pressure upon the Governor and Attorney General, the *Morning Herald* wanted to keep Lewis Emery, Jr., in the legislature instead of sending him to Congress.[25] A new rate agreement between the Standard Oil and the trunklines charging 80¢ a barrel from Olean to the seaboard with a 40¢ rebate to the Standard Oil complicated the task of closing the injunction Suits. Sherman had learned that the Governor would not permit any

[22] *History of the General Council,* pp. 29–30.

[23] *Ibid.,* pp. 26–29; Sherman Papers, MS Memorandum, May 9, 1879, Sherman to Shiras, May 21, 1879.

[24] Sherman Papers, Henry W. Palmer to Sherman, May 13, 14, 1879, Sherman to Palmer, May 13, 1879; *Titusville Morning Herald,* May 15, 16, 1879.

[25] *Titusville Morning Herald,* May 14, 1879.

injunctions injurious to the Pennsylvania Railroad's freight business but would give every assistance to securing a uniform antirebate consent injunction decree. He reported to Shiras great progress toward this goal and asked him to secure Thomas A. Scott's approval. If Shiras succeeded, Palmer would cooperate regarding the Erie Railroad and the American Transfer Company. Dividing the opposition by pitting the railroads against the Rockefeller combination was, Sherman reiterated, "the only hope of success." It would bring the railroads "better rates of freight" and free the producers from the Standard Oil monopoly. Oil refining in Pennsylvania would revive and oil exports from Philadelphia and Pittsburgh would regain their former importance. With the uniform consent decree in hand Sherman expected that "a vigorous fight against the Standard and Pipe Lines will compel them to kneel to the railroads." "Absolute *secrecy*" was essential to success of the negotiation as was also "the greatest energy and speed." The negotiations should be between attorneys. When the consent decree was agreed to Campbell and the producers' General Council should be informed.

Shiras won the Baltimore and Ohio Railroad to the proposed uniform consent injunction decree. He also won the approval of John Scott, General Solicitor of the Pennsylvania Railroad. Hampton, the Pennsylvania's chief attorney in the Suits, asked Sherman to sketch "a form of a decree that would satisfy us." Sherman had received informal approval from the attorneys of the Lake Shore, Dunkirk, and Atlantic and Great Western railroads. Only Vanderbilt's approval remained to be gained.[26] Then, inexplicably, the Supreme Court of Pennsylvania ruled that it lacked jurisdiction over the Atlantic and Great Western Railroad because of the receivership instituted by a Philadelphia municipal court situated 300 miles from that insolvent carrier's line.[27] Negotiations for the decree continued, consuming six weeks before William K. Vanderbilt returned from Europe.

Sherman briefed Palmer on the developing situation, including

[26] Sherman Papers, Sherman to Campbell, May 21, 1879, Sherman to Shiras, May 21, 1879, Shiras to Sherman, May 24, June 5, 1879.

[27] *History of the General Council,* p. 26.

the latest Standard Oil-trunkline attempt to crush the Tide Water-Reading Railroad competition. Sherman again requested permission to widen the injunction and *quo warranto* proceedings so as to heighten carrier interest in the uniform consent decree and win thereby "*equal* and *open* rates." Sherman and Shiras backed Jenks's attempt to persuade Governor Hoyt to approve of extraditing the Rockefeller brothers, Bostwick, Flagler, Girty, and Pratt for trial before Clarion County Quarter Sessions. Shiras informed Palmer of how Rockefeller and Flagler had achieved temporarily a secret net rate of 10¢ per barrel to New York harbor by a rebate coup at the expense of the trunklines. Bishop, Adams & Bishop of Cleveland supplied Sherman with secret rate contracts between the Standard Oil and the railroads.[28]

By June 30 Sherman had won assent to the uniform consent decree from all of the carriers except the New York Central, which he approached through Judge R. Brown, attorney for the Dunkirk branch line into the Oil Region. By means of the contemplated arrangement each carrier would receive from $1 to $1.25 a barrel on oil shipped to New York instead of the existing low rate resulting from the rate war. Brown then secured Vanderbilt's consent and informed Sherman accordingly on July 8.[29] Sherman stood on the verge of a decisive victory.

On that same day he advised Emery, Jr., on how to terminate the United Pipe Lines' unscrupulous tactic of leaving oil in producers' tanks after purchasing it, issuing certificates on it, and then charging speculators for storage.[30]

Sherman completed the draft of the uniform consent decree on July 10. It would enjoin each carrier from discriminating in rates or in shipment facilities on lots of 1,000 barrels or more, from combin-

[28] Sherman Papers, Sherman to Palmer, June 20, 1879, Sherman to Lyman D. Gilbert, June 26, 1879, Shiras to Sherman, June 22, 1879, J. P. Bishop to Sherman, June 22, 1879, Sherman to Campbell, June 23, 1879; *History of the General Council,* p. 30.

[29] Sherman Papers, Sherman to R. Brown, June 24, 30, 1879.

[30] *Ibid.,* Sherman to Emery, Jr., July 8, 1879.

ing or agreeing with any other corporation, firm, or person to pay rebates, drawbacks, or commissions on petroleum freight rates, and from interfering with or breaking connections with connecting roads in the petroleum traffic.[31] If generally accepted and observed, this decree would have terminated the rebate system and the exclusive service favors which had contributed so largely to the rise of the Standard Oil to ascendancy in the petroleum industry. As Sherman informed Acheson, "if it can be accomplished" this "will end the troubles of oildom, while it will give the railroads a paying rate of freight instead of the bankrupt rate now in force, which is costing them not less than $40,000 per day. . . . The time to do this (if it can be done) is *now.*" He asked Acheson to take the draft to Hampton and then urge Palmer to authorize filing the additional injunction bills immediately. Acheson promptly handed the draft decree to Hampton, who promised Shiras the Pennsylvania Railroad would accept it, after which he would take a copy to Vanderbilt himself. Sherman informed Campbell of his interview with Judge Brown and urged that he see him and take the lead in securing a general railroad acceptance. He asked Campbell also to "stir up the Governor" in behalf of the extradition of the nonresident defendants in Commonwealth *v.* Rockefeller *et al.* for the trial at Clarion.[32]

John D. Rockefeller was bitterly opposed to being extradited. He pressured the New York Governor not to consent to it and personally urged Governor Hoyt not to authorize it. At Harrisburg on June 27, Hoyt, Palmer, and Secretary of State Matthew S. Quay had heard Jenks argue that defendants were extraditable if they had committed overt acts within Pennsylvania in pursuance of the conspiracy even if the latter had been formed elsewhere. Although he had been briefed by Sherman on the strategic objective behind the conspiracy suit, Hoyt took the question of extradition under advisement when the defense counsel replied that a person was extradita-

[31] *Ibid.,* MS draft "Proposed Decree . . . July 10, 1879."

[32] *Ibid.,* Sherman to Acheson, July 11, 1879, Sherman to Campbell, July 11, 1879, Acheson to Sherman, July 14, 23, 1879.

ble only if he had fled the state where the alleged crime had been committed.[33]

After awaiting Hoyt's decision on extradition for four and a half weeks Campbell telegraphed him at Sherman's instigation to request immediate action. Campbell threatened angrily, if he did not reply, to communicate with him through the press. Deputy Attorney General Gilbert replied instantly that Palmer was ill. Campbell answered tartly that he was "tired of 'addition, division and silence,'" repeating his threat. He then sent Sherman a draft letter to the Governor that he proposed to publish in the *Pittsburgh Telegraph.* If denied the extradition, Campbell threatened to take the fight to the courts of another state. Gilbert replied: "No answer can now be given to you because no decision will be rendered until the examination now commenced is ended."[34] This referred obviously to General Sweitzer's hearings. Sherman suspected with reason that Hoyt was yielding to Standard Oil pressure. Would Hoyt delay action on extradition until after termination of General Sweitzer's hearings? That would jeopardize the attempt to disrupt the Standard Oil combination with the trunklines. Subsequently in his history of the Suits Sherman would suggest that "a post mortem examination" would "reveal the secret malady to be guarded against when a wealthy and powerful malefactor, in league with powerful corporations is sought to be brought to justice."[35]

His relations with Campbell were most cordial. The latter entertained Mrs. Sherman and "*The Baby*" at Fairfield in late July while Sherman was arguing cases before the Pennsylvania Supreme Court at Pittsburgh. Afterward Mrs. Campbell informed him that that invitation had been "one of the few sensible acts she ever knew her husband to be guilty of."

During that visit the Bradford producers at a mass meeting had adopted a declaration of rights. Campbell did his best there to

[33] *Ibid.,* MS Memorandum May 9, 1879; *History of the General Council,* pp. 34–35; Nevins, *Study in Power,* I, 313–315.

[34] Sherman Papers, Campbell to Sherman, July 30, 1879, Campbell to Hoyt, July 31, 1879, Sherman to Campbell, July 30, 31, 1879.

[35] *History of the General Council,* pp. 35–36.

persuade the producers to back this up by effective financial support of the Commonwealth Suits. The producers' decreasing interest in the Suits was attributed by D. A. Wray of Bradford to their increased dependence upon "the monopoly," which deprived them of "the courage and manhood . . . to enable them to strike a blow for liberty." [36] The great decline in crude-oil prices arising from the Bradford glut had deprived many of the sinews of war, however.

Sherman understood. He advised Campbell to withhold his open letter to Hoyt until it could be ascertained "what can be done." Sherman conceded that the plight of the impoverished producers deprived them of "the means to prosecute the suits so auspiciously begun, and successfully carried on. Nor do they seem to have the disposition." Yet, no litigation of such magnitude had ever before in Pennsylvania been "prosecuted so vigorously and at so little expense in money." However, without "immediate contributions of money for necessary expenses" the Commonwealth Suits would have to be abandoned. The Associate Counsel were not asking for their fees "at present," he added.

He interpreted Gilbert's ruling to mean that Hoyt would not decide the question of extradition until the civil Suits were disposed of "or at least until the testimony is taken in them." Such inaction, Sherman asserted, "is of course a substantial denial of justice." Quixotically, Palmer had already stopped proceedings against the United Pipe Lines "until the *Criminal prosecution* was ended." Palmer's contradictory policy Sherman termed a "gross outrage," whose "insolent purpose" was clearly to contrive so that "each prosecution stops the other," since according to Palmer the criminal-conspiracy defendants need not testify in the civil suits. During the delay "Rockefeller & Co. are to be allowed the summer and fall in which to rob the producing interest 'with their gloves off' while the State authorities chain up the dogs, and muzzle them until the outrage is consummated." As senior Associate Counsel, Sherman recommended "a final effort" that would place "the responsibility for failure—if failure must come . . . where it belongs."

[36] Sherman Papers, Campbell to Sherman, July 30, 1879, D. A. Wray to Campbell, July 30, 1879.

Campbell's duty, he insisted, was "to exhaust every resource & take every path, before giving up the fight." He should appoint "a few reliable men" to raise necessary funds from the "Bradford producers who are the worst sufferers. . . . Some one should . . . ascertain whether there is any probability that the Penna. R.R. will agree to a compromise. If they will not, that branch of the matter is at an end" since the courts would "never enjoin it alone." He continued: "Unless assisted by the State authorities and without money we cannot reach the other roads, except by agreement." The entire subject, he felt, should be reported and thoroughly discussed secretly at the next General Council on August 12. *"The result should not be made public at once."* With funds in hand the Suits could proceed. If these efforts failed, "there is nothing left but to prosecute criminally in the State of New York, and this can be left for future considerations." If the fight had to be given up it should be made clear that "it is no fault of ours or of the Council but the responsibility rests mainly upon the state authorities," and the "residue upon the inability of the producers to carry on the legal proceedings necessary to enforce the law." [37] Sherman was ready to fight to the end if money sufficient for court expenses was forthcoming.

When Daniel O'Day gave bail at Clarion Quarter Sessions on August 1, Sherman vetoed publication of Campbell's open letter to Governor Hoyt. Ezra Crossman, a Pittsburgh and Bradford attorney, had discovered evidence that would enable the Commonwealth to prove that the New York, Lake Erie & Western Railroad had contracted to provide the Standard Oil with exclusive tank car service. Confidential information indicated that at a Saratoga conference the trunkline officials were disposed "to throw the Standard overboard if this can be done." However, Shiras's behind-the-scenes negotiations with that end in view were abortive. After Sherman devoted August to the Tide Water, Equitable, and Ohlen litigation, he asked Campbell for information on the status of the Common-

[37] *Ibid.*, Sherman to Campbell, July 31, 1879.

wealth Suits. He wished to know how much time he must spend on them "so that I can arrange other business." [38]

Shiras said that Hampton and John Scott desired to settle their injunction case with the Commonwealth "but I fear they will be unable to get a concert of action." Shiras consented to another week's postponement of Sweitzer's hearings while Hampton tried to persuade Thomas "Scott and others to a final decision, as to the proposed decree." By this time Palmer intimated that Hoyt would refuse extradition in Commonwealth *v.* Rockefeller *et al.* and that he approved. Campbell then wrote Sherman: "There is nothing for me but to fight it through & I think it can be done more speedily than we have counted on." He asked Shiras to schedule Commonwealth *v.* Pennsylvania Railroad for hearing by the Pennsylvania Supreme Court.[39]

The diminished prospect of success in Commonwealth *v.* Rockefeller *et al.* can be traced to Matthew S. Quay, rising star in Senator Cameron's organization. Quay informed influential persons, Sherman learned, that Hoyt would delay decision upon extradition until after trial of the intrastate defendants. If they were convicted he would extradite the out-of-state defendants, "otherwise he will not." [40] Hampton then asked for further delay of Sweitzer's hearings so as to enable the trunkline executives to act upon the proposed consent decree. Despite Sherman's and Shiras's objections Sweitzer postponed his hearings until December 10, after the end of the pending Supreme Court term, which he set as the time when the Pennsylvania Railroad must close testimony. After that Shiras informed Sherman that he had "no confidence left in anything that Mr. H. may say." [41] Then John D. Archbold as Rockefeller's repre-

[38] *Ibid.,* Campbell to Sherman, August 1, 3, 1879, Sherman to Campbell, August 1, 29, 1879, Sherman to Shiras, August 18, 1879, E. Crossman to Crawford and Campbell, August 18, 1879.

[39] *Ibid.,* Shiras to Sherman, September 2, 1879, Campbell to Sherman, September 3, 1879.

[40] *Ibid.,* Sherman to Campbell, September 8, 1879.

[41] *Ibid.,* Sherman to Shiras, September 15, 1879, Campbell to Sherman, September 15, 1879, Shiras to Sherman, September 15, 17, 1879.

sentative proposed to Sherman a negotiated settlement of Commonwealth *v.* Rockefeller *et al.* With Campbell's consent Sherman listened to the proposition. He consented to postponement until mid-December of the O'Day portion of Commonwealth *v.* Rockefeller *et al.*

Campbell regarded Archbold's proposition as indicative of "how badly they are scared." He predicted that the Standard Oil men's "zeal for a settlement will probably play out." He reported to Sherman that "Jenks is very confident of conviction when the case comes to trial." Meanwhile, the *Oil City Derrick* attacked Sherman and the General Council in an abusive article. Sherman prepared the bill of particulars for Commonwealth *v.* Rockefeller *et al.*, drawing upon his unequalled knowledge of the history of the petroleum industry, pipeline, and railroad interrelationships. After Clarion Quarter Sessions upheld his bill, trial was set for the third Monday in December. Sherman also drafted an astute memorandum on the corporate history and economic functions of the American Transfer Company for the contemplated *quo warranto* proceeding against it.[42] Meanwhile, the producers movement languished. Only twelve representatives including Campbell and Patterson attended the November 11 General Council meeting for the election of officers. However, Acheson believed that Jenks's case against the intrastate defendants in Commonwealth *v.* Rockefeller *et al.* was "impregnable." Acheson and Campbell pledged Sweitzer to hold the Pennsylvania Railroad to its closing date of December 10.[43]

The vulnerability of the defendants in Commonwealth *v.* Rockefeller *et al.* made Standard Oil headquarters increasingly nervous as the trial date approached. Archbold telegraphed an invitation to Sherman and his principals to "meet us in New York Monday for the purpose of settling everything." John Scott informed Shiras and

[42] *Ibid.*, Archbold to Sherman, October 1, 10, 1879, MS Memorandum, October 13, 1879, Sherman to Wilson & Jenks, October 2, November 17, 1879, Sherman to Archbold, October 13, 1879, MS Agreement between Sherman and S. C. T. Dodd, October 2, 1879, Campbell to Sherman, October 7, November 16, 1879.

[43] *Ibid.*, Campbell to Sherman, November 16, 17, 1879; *Titusville Morning Herald*, November 12, 1879.

Campbell that the Standard Oil chiefs "were afraid the excitement of the trial might be fatal to both Lockhard & L. A. Scott." Shiras expected that "the result of the trial would be to put all the defendants behind the bar of the penitentiary." Campbell and Sherman agreed that to go to New York for such a conference would "merely inspire" the Standard Oil men "with fresh confidence and put us in a fatally false position." Campbell suggested that Sherman join him at Philadelphia within a few days and inform Archbold that they could see him there secretly "Thursday morning" at the Girard Hotel. While a defense counsel predicted that Commonwealth *v.* Rockefeller *et al.* would never be tried, Campbell telegraphed Sherman in cypher, "Mellon wants meeting Naughtiness Monday." [44]

Sherman did not join him in Philadelphia. Franklin B. Gowen warned that any such conference should be between counsel lest abortive negotiations be brought out in the trial to the injury of the prosecution in Commonwealth *v.* Rockefeller *et al.* Benson and Satterfield vetoed the proposed New York meeting. Sherman informed Archbold, who renewed the invitation to a conference, that "there is no time now to be spent in useless talk . . . if any propositions are to be made they should be made through counsel." Tide Water headquarters urged that he push ahead with the Commonwealth Suits. Sherman agreed that any conferences "would end in wind." [45]

Campbell persuaded Colonel Joseph D. Potts of Philadelphia, an active backer of the Tide Water, to testify at the Clarion conspiracy trial. Guided by Potts Campbell subpoenaed Thomas A. Scott to produce "some letters that will astonish you," he informed Sherman. Campbell agreed that Sherman had acted wisely in reply to Archbold's overtures. After asking Sherman to draft a statement of expenses incurred "& possible liability for costs," Campbell added,

[44] Sherman Papers, Campbell to Sherman, November 29, 30, 1879, S. P. Franchet to B. D. Benson, November 29, 1879, Sherman to Campbell, December 2, 1879, J. D. Archbold to Southard & Barstow, November 29, 1879.

[45] *Ibid.*, Sherman to Campbell, December 2, 1879.

"if any more propositions are made you & I can consider them then but I feel my confidence in the case growing sufficiently to demand as a sina qua non the abandonment of all rebates by the railroads & a contract between them & myself & all others" securing to the independents oil rates as low as those charged others and not to exceed $1 per barrel. This would protect the Tide Water "especially if immediate shipment is also abolished. Lockhart is still sick & Warden told my friend P that he was extraordinarily alarmed."[46]

As December 15 approached, defense motions for a continuance of Commonwealth *v.* Rockefeller *et al.* were overruled. Great popular interest in the case contrasted with the little attention paid to it in the oil trade. The uncertain outcome, the common-law basis of the indictment, and the basis for individual damage suits that convictions would establish were all described in the press. In preparation for the trial Sherman obtained the brief filed by Bishop, Adams & Bishop in Scofield, Shurmer & Teagle *v.* Lake Shore and Michigan Southern Railroad Company, which had exposed that carrier's rebates and service favors to the Standard Oil.[47] Commonwealth *v.* Rockefeller *et al.* promised to be the most spectacular, popular, and possibly the most important of the Commonwealth Suits.

Meanwhile, General Sweitzer resumed the Titusville hearings at which the Pennsylvania had been ordered to close its testimony. There Hampton presented to the Master a letter from Governor Hoyt saying that Shiras had been instructed by Palmer that the defendants in the five equity injunction proceedings were not required to close testimony individually until the Commonwealth had closed its testimony against all of them.[48] This method of announcing the Governor's decision was an insult to Sherman, the senior Commonwealth Associate Counsel. It created an almost insuperable obstacle to completion of the Pennsylvania Railroad and United Pipe Lines cases. It prevented early reference of the case against the Pennsylvania Railroad Company back to the Supreme Court for

[46] *Ibid.,* Campbell to Sherman, December 5, 1879.

[47] *Ibid.,* Campbell to Sherman, December 5, 1879, J. P. Bishop to Sherman, December 6, 1879.

[48] *Ibid.,* Sweitzer to Sherman, December 7, 1879; *History of the General Council,* p. 25.

adjudication of the legal issues involved. In this manner Hoyt informed Sherman and Shiras that he had switched his confidence from them to the Pennsylvania Railroad.

Hoyt's action diminished the pressure upon all the railroads involved in the Suits to agree to the proposed uniform consent decree. The probability that the Associate Counsel would be able to complete their cases against the Dunkirk and Lake Shore railroads was remote because their officials were beyond Pennsylvania judicial process. Because of the Supreme Court's ruling, the Atlantic and Great Western Railroad's receiver could not be reached. The *quo warranto* against the United Pipe Lines had been allowed to sleep. It was now improbable that Palmer would authorize extension of the injunction proceedings to include the reorganized Erie Railroad and a *quo warranto* against the American Transfer Company. Thus, all the weapons that Sherman had fashioned to win general abolition of rebates and service favors on the railroads and pipelines were struck from his hands as the result of the pressure that the defendants, chiefly the Pennsylvania Railroad and the Rockefellers, had brought to bear upon Governor Hoyt. Sherman, Campbell, the independent producers, and the independent refiners all believed now that they had been betrayed by Hoyt despite his and Palmer's solemn promises during the political campaign of 1878.[49]

There remained to be ascertained what could be accomplished in Commonwealth *v.* Rockefeller *et al.* at Clarion on December 15. Sherman and Jenks were confident of their ability to win convictions. Were they to succeed they would have developed a method in criminal law universally applicable to business enterprisers entering into conspiracies against their competitors with the railroads. Sherman viewed this case properly therefore as "one of the very first importance." In preliminary consultation he assigned to Jenks the actual trying of the case assisted by two other attorneys retained by the Commonwealth at the expense of the General Council of the Producers Unions. Sherman attended in a general supervisory capacity but did not expect to participate directly in the trial of the case.

[49] *History of the General Council,* pp. 25–29.

He would bring to Clarion for the prosecution's use all the literature of the producers' movement and the testimony taken in the Civil Suits.[50]

He was unprepared for the defense's application on December 11 before Supreme Court Associate Justice Edward U. Paxson for a writ of certiorari in behalf of Vandergrift, Lockhart, Warden, and O'Day removing the case and its records to the Supreme Court on the ground that a fair trial could not be held in Clarion County because the judge and prosecution were allegedly members of or interested in the Producers Unions. Defense counsel cited also the "precedency" and incomplete taking of testimony in the civil Commonwealth Suits, despite the Governor's order that the defendants in the conspiracy case be not required to testify in them before their own trials. Although the District Attorney had not been notified of the application, Paxson issued a rule ordering the prosecution to show cause on January 5, 1880, why the writ of certiorari should not issue.[51]

Clarion was anticipating the trial of the Rockefeller coterie with great satisfaction, partially because of the crowd of visitors that it would attract. Since Rockefeller and his lieutenants were unpopular in Clarion County because of their monopolistic practices it might have been difficult to choose an impartial jury to try them. Now, it appeared that the producers' desire for vengeance upon the powerful, pitiless monopolists was in danger of being frustrated.

The *Morning Herald* remarked that since an appeal from the trial court was certain after conviction, the sooner the issue was taken to the Supreme Court for adjudication of the principles involved in the transportation problem that was troubling so many state legislatures and Congress the better. "What we want," the *Morning Herald* declared, "is substantial business results and the final settlement on imperishable foundations of great commercial

[50] Sherman Papers, Sherman to Jenks, December 9, 10, 1897, Campbell to Sherman, December 12, 1879.

[51] *Ibid.*, MS copy of Supreme Court record of rule, December 11, 1879, Sherman to Campbell, December 9, 1879, Sherman to Jenks, December 10, 1879, Campbell to Sherman, December 12, 1879; *Titusville Morning Herald*, December 15, 1879; *History of the General Council*, pp. 36–37.

principles." Paxson's rule necessitated postponement of the trial date of Commonwealth *v.* Rockefeller *et al.* until January 19. Meanwhile, the *Philadelphia Daily Times,* organ of the "Pennsylvania ring," the Standard Oil-controlled *Bradford Era* and *Oil City Derrick,* the *Railway World,* and *Morning Herald* published articles favoring the defendants which Sherman believed were intended to influence the public and the court.[52]

Paxson's action was extraordinary, and Sherman knew it was.[53] The Pennsylvania Constitution of 1873 had abolished the Supreme Court's power to remove cases to it for trial by writ of certiorari and had granted it appellate jurisdiction only! Paxson's rule violated that section of the Constitution. Was he attempting thereby to regain the lost portion of the Court's former jurisdiction? Was the defense counsel in Commonwealth *v.* Rockefeller *et al.,* knowing that such an appeal from the Constitution of 1873 to the Constitution of 1838 which it had superseded could hardly be upheld by the full bench of the Supreme Court, resorting to Fabian tactics in applying for the writ in an attempt to defeat the Commonwealth? The prosecution in this as in the civil Commonwealth Suits was financed by the impoverished, increasingly discouraged producers. The strategy of delay might make more appealing Archbold's recent proposition to Sherman that their principals meet to negotiate a settlement of all of the Commonwealth Suits.

Sherman prepared a paperbook containing the prosecution's argument and a review of the facts of Commonwealth *v.* Rockefeller *et al.* for presentation to Judge Paxson at Philadelphia. Simultaneously, he rallied his friends to support the successful attempt to have Mark W. Acheson, one of his Associate Counsel, appointed United States District Judge at Pittsburgh by President Rutherford B. Hayes. He thanked Simon Sterne of New York City for copies of the testimony taken by the Hepburn Committee at Albany in its investigation of the railroads of New York and requested copies of the arguments of counsel before it. He was advised privately that

[52] *History of the General Council,* pp. 38–39; *Titusville Morning Herald,* December 16, 1879.

[53] Sherman Papers, W. L. Corbett to Sherman, December 20, 1879, Sherman to Corbett, December 19, 1879.

the application for a certiorari would be denied the defense in Commonwealth *v.* Rockefeller *et al.*[54]

Standard Oil headquarters in New York City reached the same conclusion. On the evening of December 22 Sherman's friend, the attorney Charles Heydrick, called at his Titusville residence and discussed with him until midnight a proposition in behalf of the defendants to settle the Commonwealth Suits. From this proposed general settlement the Tide Water was to be omitted. Sherman then drafted "a memorandum of vital points to be covered in the event" that Heydrick would "present any proposition worthy of consideration."[55]

Neither Sherman nor Campbell could refuse to consider such a proposal. The producers' condition had deteriorated steadily since the preceding spring. Oil was now well below $1 a barrel. The United Pipe Lines was attempting to collect 20¢ a barrel for pipage from producers at a time when it was charging the Standard Oil 5¢, after reviving "immediate shipment" as part of Rockefeller's policy of handling the independent producers' opposition "without gloves." The relief brought them by the Tide Water's success was now offset by the low price of oil. An increasing number of producers were reduced to virtual poverty as the business depression neared its end.[56]

The Oil Region's resentment at the brutal tactics of the Standard Oil and the United Pipe Lines was undiminished, as indicated by the spasmodic demonstrations in front of their Bradford offices, and by the processions of masked men that hooted as they passed before these offices and those of pro-Standard-Oil newspapers at Bradford and elsewhere. The few remaining independent newspapers advised the producers to resort to force to wrest a redress of grievances from the oil combination, although President Campbell rejected such a policy.[57]

[54] *Ibid.,* Jenks to Sherman, December 22, 1879, Campbell to Sherman, December 22, 1879, Sherman to Sterne, December 17, 1879.

[55] *Ibid.,* MS Memorandum December 23, 1879, Sherman to Campbell, December 23, 1879.

[56] *History of the General Council,* pp. 30–31.

[57] *Ibid.,* p. 31.

# CHAPTER VIII

# *A Negotiated Settlement*

THE great struggle between the independent oil men and the Standard Oil combination had developed into a war of attrition. However, the extraordinary disclosures of the character and methods of the great oil monopoly made during General Sweitzer's hearings and the parallel Hepburn Committee's investigation had attracted widespread attention. The influential but distant *Chicago Tribune,* for example, editorialized stingingly upon Rockefeller's methods and the implications of his monopoly. Such newspapers attacked his methods and those of his railroad allies, reviving popular antimonopolist stereotypes, while demanding national railroad regulation, a proposal which the New York Chamber of Commerce was supporting after its recommendation by the Committee on Legislation of the General Council of the Petroleum Producers Unions.[1]

The interminable delays that had prevented bringing the Commonwealth Suits before the Supreme Court of Pennsylvania for adjudication undermined the producers' confidence in their efficacy as a remedy for bitter grievances. As they were impoverished they became convinced that the oil combination was unafraid of the Suits, and that prosecution of these had only incited it to doubled exactions and more ruthless methods. Confidence in Governor Hoyt waned steadily, as leaks from the Committee on Legal Remedies told producers how he and Palmer yielded to pressure from the defendants.[2] All this diminished the Producers Unions' ability to finance the Suits.

[1] *Chicago Tribune,* November 25, 1879, January 28, February 20, 1880; Lee Benson, *Merchants, Farmers & Railroads* (Cambridge, 1955), pp. 115–132.

[2] *History of the General Council,* p. 32.

Yet, by late December 1879 Sherman and his Associate Counsel had made their case before the general public. It became generally recognized that the Suits had disclosed the true character of the Standard Oil combination. The success of the Tide Water had broken its monopoly of transportation to the Atlantic coast and revived competition in oil buying in the Oil Region. The railroad trunklines were still suffering heavily from the oil rate war against their Tide Water-Reading Railroad rivals. They set an example that led the *Railway Gazette* to predict that long-distance pipelines would soon deprive the former of their crude-oil traffic.[3]

The attempts of the Producers Unions to curtail production and raise the price of crude oil had failed. Their program of producer-owned storage was frustrated by the hostile United Pipe Lines Company. The Standard Oil was buying crude at unprecedentedly low prices and paying proportionately low freight rates during the rate war. Thus its profit position was actually enhanced during the struggle rather than impaired because of the exposures of its lawless and unethical methods. However, the personal reputations of the Rockefellers and their chief lieutenants were besmirched further by the revelations of the Hepburn inquiry. Outraged public opinion in New York and Pennsylvania was demanding remedial legislation that would weaken the Standard Oil's hold upon the railroads and curtail its transportation privileges.

Although Governor Hoyt and Attorney General Palmer were changing sides in the Commonwealth Suits, and desired their settlement in a manner that would not seriously antagonize or injure the powerful defendants, John D. Rockefeller's fear of extradition and trial in the conspiracy case was such as to offset partially his stronger strategic position. Both sides to the great contest had suffered sufficiently to be ready to consider a negotiated settlement such as Quay had urged upon Campbell in January 1879 and Archbold had proposed to Sherman in December.

Heydrick negotiated with Sherman for two days in behalf of the Standard Oil and its pipeline subsidiaries. This renewal of Arch-

[3] Quoted in Johnson, *op. cit.*, p. 96.

bold's efforts indicated that Rockefeller preferred a negotiated peace to legal war. Sherman's leadership in the Commonwealth Suits and his relation to the Tide Water and Ohlen had secured for the independent oil men the status of a first-class business force. He believed that he had won the campaign, but not the final battle. This was indicated by a memorandum embodying points agreed upon that he handed to Heydrick on the latter's invitation on December 23. It dealt with all aspects of the producers' problems except trunkline rates and service policies.

Sherman proposed a formal contract between representatives of the General Council and members of the Producers Unions signing it on the one hand and the Standard Oil companies of Ohio and Pennsylvania and the corporations "acting in harmony" with them (including the United Pipe Lines and American Transfer companies) on the other hand. This contract should specify abandonment by the parties of the second part "of all forms of discrimination, preference, and favoritism" in oil prices, basing price instead upon tested quality only, the abandonment of "immediate shipment" and rebates on pipeline transportation, adoption of uniform, reasonable, contracted pipage and storage rates, use of excess lower-district tankage for storage of surplus upper-district oil, observance of expiration dates of oil certificates, and adoption of the independent tank-owners' specifications for pipeline service. These covered "most of the pressing evils under which the producers labor" although "other matters" that should be included in the contemplated agreement would doubtless occur to them both, he informed Heydrick. The document was suggestive only and strictly confidential, he added.[4]

Its contents attested to the strength of Sherman's position and to his skill as a negotiator. A redress of grievances by the oil combination was to be the price paid for abandoning the conspiracy case and the civil Commonwealth Suits against the United Pipe Lines. From this it can be seen that, in a manner not initially anticipated,

[4] Sherman Papers, Sherman to Heydrick, MS Memorandum, December 23, 1879.

the impending criminal prosecution had impelled the frightened Rockefeller to offer terms. What Sherman sought, it became evident, was an end to the more obvious abuses perpetrated by the oil monopoly in the conduct of its business instead of venting personal spleen upon its head by pillorying Rockefeller in court before the public. Sherman had not entirely given up hope for the negotiated consent decree applicable to the railroads. That alone held out hope of destroying the privileged basis of the monopolistic Rockefeller organization.

The United Pipe Lines announced suspension of "immediate shipment" the following day.[5] Sherman informed Campbell of Heydrick's overtures and that he had not included the Tide Water, but he did not enclose a copy of the memorandum. Sherman suggested that they confer in Philadelphia, where he would argue before Justice Paxson against the proposed certiorari. By this time Shiras had withdrawn from active participation in the Commonwealth Suits, most of which were in abeyance as the result of Governor Hoyt's and Palmer's decision to delay extradition and the questioning of indicted defendants before General Sweitzer. Campbell lacked confidence in the Standard Oil men's good faith, although he conceded Heydrick "is the best & fairest ambassador they have yet sent." Campbell believed that they should compel the railroads to accept a consent decree in the injunction suits, as the best and "surest" mode of settlement. He insisted that the railroads "are ready to settle." [6]

Before Sherman left Titusville to confer with him in Philadelphia at the Girard Hotel, whose table and "the pervading Democratic atmosphere" he found congenial, Heydrick asked him to prepare a draft contract embodying his "views" but omitting overproduction, the price of oil, "and the like." Heydrick added:

[5] *History of the General Council,* pp. 39–40; *Derrick's Handbook,* p. 320.

[6] Sherman Papers, Campbell to Sherman, December 24, 25, 1879, Sherman to Campbell, December 26, 1879; Lloyd Papers (State Historical Society of Wisconsin), Shiras to Henry D. Lloyd, December 26, 1879.

My theory has always been that if you can confine the common carriers to their legitimate business and compel them to discharge their duties to the public—viz to furnish reasonable facilities for transportation and carry for all on equal terms and at rates that shall not be changed except upon reasonable notice, you will eradicate all evils except over production. I expect to meet you again soon to come to a definite conclusion.

Sherman conferred with George Lear on his way to Philadelphia, and agreed with Campbell there that the negotiation with Heydrick should continue.[7]

Upon returning to Titusville Sherman received a request from Henry Demarest Lloyd, financial editor of the *Chicago Tribune,* for a copy of the printed testimony in Commonwealth *v.* Pennsylvania Railroad Company. Lloyd used this and pamphlets that Sherman sent, and Patterson's analysis of railroad favoritism, in preparation of editorials, i. e., "A Corner on Light," and of his famous *Atlantic Monthly* article, "The Story of a Great Monopoly." These articles acquainted the general reading public with the extraordinary disclosures made by the Commonwealth Suits and by the Hepburn Committee of the Standard Oil's privileged position on the railroads while describing the producers' war on that great monopoly.[8] Sherman's compliance with other Lloyd requests for data and references stimulated that publicist to broaden his antimonopoly journalistic crusade from the railroads to include industrial monopoly. This campaign made the *Chicago Tribune* an extremely influential factor in the enlightenment of the general public and the competitive business community on the monopoly problem. The two men became warm friends.

At a three-day conference in New York City from January 7 to January 9, 1880, Sherman and Heydrick were joined by committees

[7] Sherman Papers, G. Lear to Sherman, December 26, 27, 28, 1879, Heydrick to Sherman, December 26, 1879.

[8] *Ibid.,* Sherman to Lloyd, January 2, March 15, July 13, 1880; Henry Demarest Lloyd Papers (Winnetka, courtesy of Mrs. Wm. B. Lloyd), E .G. Patterson to Lloyd, January 31, 1880; *Chicago Tribune,* November 25, 1879, January 28, February 20, December 31, 1880, February 22, 1881.

representing the opposing sides. The second suspension of "immediate shipment" by the United Pipe Lines had softened asperities and made the negotiation possible.[9] On the day before the conference began State Senator W. L. Corbett had presented to Justice Paxson Sherman's argument that the Pennsylvania Constitution of 1873 had abolished the Supreme Court's original jurisdiction in criminal cases, which prohibited that Court from trying Commonwealth *v.* Rockefeller *et al.* should its record be removed to it for this purpose. After the defense counsel argued that doing so would not be an exercise of original jurisdiction, the Court ordered that the matter be reargued before its full bench and stayed proceedings on the case at Clarion until then. This delay, when the order for reargument was announced on January 7, weakened Sherman's position in the secret negotiation and diminished his hope of victory in the criminal case, now the key proceeding of the Commonwealth Suits.[10]

On January 7 and 9 at the Fifth Avenue Hotel Heydrick had offered to Sherman the Standard Oil's terms. These included a contract renunciation of rebates and drawbacks on the railroads except those they were at liberty to give shippers, basing crude-oil prices on quality, equalization of certificate and "immediate shipment" oil prices, the purchase and piping of all oil produced up to 65,000 barrels per day for fifteen consecutive days, and payment up to $40,000 of expenses incurred by the producers in financing the Commonwealth Suits. The Standard Oil's written proposition embodying these proposals was agreeable to the representatives of the independent oil men except for the Standard Oil's reservation of the right to accept any rebate the railroads *"were at liberty to give to other shippers."* Sherman demanded instead of the proposed unenforceable contract between the Standard Oil and the independents enforceable court decrees in the existing Suits. He attempted vainly to persuade the Standard Oil to oblige the railroads to reform their oil-rate policies. Heydrick insisted that decrees against the railway companies must be obtained by means of the Suits against them and

[9] Sherman Papers, "C. H." to "Fd Ho" (twice), January 7, 1880.

[10] *Titusville Morning Herald,* January 6, 9 (quoting *Railway World*), 14, 1880; *History of the General Council,* p. 39.

by amicable action and consent, but promised that the Standard Oil would pledge itself to consent to any arrangement made for that purpose. The Standard Oil-controlled pipelines, he promised, would not issue excess certificates. The United Pipe Lines would not lower pipage and storage rates, as Sherman demanded. On January 9 Heydrick asserted that his clients had agreed to all that Sherman asked except joint action to secure the railroad consent decree and reduction of the United Pipe Lines' rates. The latter Heydrick declared would be unfair to the producers and oil shippers.[11]

Sherman returned to the charge on January 15 when he demanded realistically that since the Standard Oil had rate and service trunkline contracts from which derived the grievances the independents complained of, the former "should see that the railroads sign the agreement" and that it should also join with the Producers Unions in securing the consent decrees prohibiting discriminatory railroad policies. One who was "willing to make a contract and intends to perform it, should not be unwilling to so bind himself that it can be carried out with practical effect in all its parts," he argued. He objected again to Heydrick's written stipulation of January 7 that his clients be free to receive any rebate or drawback that the railroads were at liberty to give to other shippers because that recognized "the lawfulness of that system which we believe to be unlawful." The settlement of the Suits should correspond with the general proposition agreed upon, i. e., the "entire abrogation of the system of rebates drawbacks and secret rates of freight." This was the Producers Unions' great objective. However, Heydrick's stipulation rested upon the Constitution of 1873 as he well knew. Sherman waived his opposition to sudden large decreases in freight and pipage rates.[12] Both attorneys, during the New York conference and in the subsequent negotiations, were engaged in the highest level of corporation law practice, representing important business interests in intricate negotiation for the effectuation of their purposes. For the Commonwealth of Pennsylvania Sherman was doing

[11] Sherman Papers, C. Heydrick to Sherman, January 10, 1879; *History of the General Council,* pp. 40–41; Johnson, *op. cit.,* p. 97.

[12] Sherman Papers, Sherman to Heydrick, January 15, 1880.

what he could to salvage victory from the hamstrung civil and criminal Suits.

Sherman then asked Judge Brown if he could meet him in New York to confer upon the subject of "petroleum freights," an obvious attempt to revive and combine the negotiation for a uniform railroad consent decree with the negotiation with the Standard Oil. Campbell then telegraphed Sherman on January 17: "Will receive answer Monday over Railroad wires parties desirous for general peace." Two days later Sherman informed Heydrick that his clients would insist upon some collateral conditions.[13] The situation developed rapidly thereafter.

Governor Hoyt supported the negotiated settlement vigorously. He told Campbell to capitalize upon this since he was going to compel one.[14] This produced a settlement within less than a month, much of which time was expended in cipher telegraphing and correspondence between the various parties and their counsel. Sentiment among the defendants developed early for substitution of a general agreement for the proposed consent decree. In ultrasecret negotiations with William G. Warden, a leading Rockefeller lieutenant, Sherman proposed inclusion of payment by the Standard Oil of the expenses incurred by the Producers Unions, their officers and members, and of the fees of the counsel they supplied to the Commonwealth for the Suits. When Warden objected, Rockefeller ordered him to agree. In late January the completed agreement was ordered engrossed, to be signed "by all" in Philadelphia and then sent to Cleveland, New York, Pittsburgh, and Oil City "for completion, when, with the annotated original" it would be sent to Sherman at Titusville "to see that it was correct." B. B. Campbell would then sign it in behalf of the Producers Unions. In the interim the expense money was to be deposited with an intermediary.[15]

[13] *Ibid.,* Sherman to Brown, January 16, 1880, Campbell to Sherman, January 17, 1880, Sherman to Heydrick, January 19, 1880.

[14] *Ibid.,* C. N. Clark to Sherman, February 28, 1880.

[15] *Ibid.,* Sherman to Brown, January 25, 1880, Sherman to Heydrick, January 26, 1880, Heydrick to Sherman, February 5, 1880, W. G. Warden to Campbell, January 27, 1880, R. H. Wallace to Sherman, January 30, 1880, Sherman to Campbell, February 2, 6, 1880.

Meanwhile, the Governor's and Attorney General's approval of the terms of the agreement had to be secured, and also their cooperation in terminating the equity and *quo warranto* proceedings and the case of Commonwealth *v.* Rockefeller *et al.* For this, Shiras's assistance was invoked. A formal agreement was reached finally between the parties on January 28 at Philadelphia. The contract between Campbell and the Standard Oil combination was signed the next day and sealed on February 5.

Adhesion of the Pennsylvania Railroad to this contract was then sought by Campbell and Sherman in a conference "with the enemy" at Philadelphia on February 10, while Governor Hoyt and Attorney General Palmer put up in the Intercontinental Hotel with the trunklines' representatives. The meeting produced an agreement to withdraw the injunction suit against the Pennsylvania Railroad. Persuasion of the Vanderbilt and Erie-Atlantic and Great Western systems to agree to abandonment of rebating was then undertaken.[16]

Until this time the meetings, journeyings, and telegraphing of the representatives of the independent oil men and of the Standard Oil and the trunklines had been kept secret. However, the *Bradford Star* published a dispatch from Foxburg on February 6 describing a secret conference between Campbell's committee and Vandergrift's committee that had been held there "looking toward a compromise of the pending suits," and also Campbell's solitary round trip on a chartered special train to Clarion together with his *sub rosa* admission that settlement of the Suits was on foot. The settlement, he said, would be accomplished on terms "honorable to all concerned." The *Titusville Morning Herald* republished this the next day. The secret was out.[17] Cipher telegrams kept Campbell, Sherman, and Shiras in touch with each other during the remaining nine days' negotiations. On February 9 Sherman hoped that John Satterfield, a large-scale producer associated with the Rockefellers in Taylor & Company, but who had supplied invaluable "inside" data to the

[16] *Ibid.,* Sherman to G. A. Jenks, February 2, 1880, Sherman to W. L. Corbett, February 2, 1880, Campbell to Sherman, February 2, 6, 9, 1880, Sherman to Campbell, February 2, 1880.

[17] *Titusville Morning Herald,* February 7, 1880.

Commonwealth during the Suits, would join him on the Philadelphia & Erie train to Philadelphia. "It is so deuced lonesome," Sherman explained.[18]

On February 12 the *Morning Herald* dignified rumors of a compromise settlement of the Commonwealth Suits with an editorial. Campbell, it observed, would be accused perhaps of selling out his constituency. His character was safeguard against that, it declared. In view of the great expense of "an aggressive war against the Standard," and the "glorious uncertainty of the law," if the producers were reimbursed in the settlement and their grievances removed then "Mr. Campbell has not been President in vain." The settlement, the editor added, would be laid before the General Council "here today" where Campbell would "probably receive a vote of thanks." The Council approved of Campbell's course. Four days later he telegraphed Sherman: "Everything closed see you Wednesday morning."[19]

Interestingly, on February 12 also it was rumored in New York City that the oil rate war against the Tide Water-Reading had been terminated there by a pooling agreement between that independent alliance, the trunklines, and the Standard Oil. New York, Lake Erie & Western Railroad officials predicted that their share of the "heavy gains" from this arrangement would increase receipts by $1,250,000 per annum. The new pool was obviously part of the settlement of the great contest between the independent producers and the oil combination.[20]

Sherman notified Judge Brown, Vanderbilt's representative, on February 16 that "under the arrangements made in reference to the discrimination Standard cases, only the bills against the Pen R.R. and U. Pipe Lines have been dismissed, and this was because a contract with them was entered into by which the rebate system was abandoned and other grievances of the petroleum trade remedied."

[18] Sherman Papers, Sherman to J. Satterfield, February 9, 1880, Campbell to Sherman, February 12, 1880, Shiras to Sherman, February 12, 1880.

[19] *Ibid.*, Campbell to Sherman, February 16, 1880.

[20] *Titusville Morning Herald,* February 13, 1880; Nevins, *Study in Power,* I, 352.

The other Suits, he said, "will not be dismissed unless the other different companies choose to make similar contracts." He added significantly: "I have understood that they did so intend, but have had no direct negotiations with them." A similar overture to the Erie was equally unsuccessful.[21]

Campbell presented the terms of the settlement with the Standard Oil and its subsidiaries and with the Pennsylvania Railroad formally to the General Council of the Producers Unions at Producers Hall, Titusville, on February 18. In anticipation Sherman had drafted a statement listing the producers' achievements in the settlement. These, he indicated, far outweighed the concessions that Campbell and he had had to make to secure them. Succinct in form, the statement was his justification for having negotiated the settlement.

I.

MEMORANDA.

*Feb. 18th*

*Local Trade in Oil Region.—Standard Contract.*

1. Abandonment of immediate shipment.
2. Receiving transportation & delivery of crude from wells, assured.
3. Uniform & reasonable rates of pipage which are not to be advanced except upon 30 days notice.
4. No discrimination in transportation or storage by pipelines between patrons.
5. Issuing of certificates for oil on demand.
6. No discrimination between districts except as to quality.
7. Positive agreement to take care of 65,000 bbls.
8. Also more than that if capacity admits.
9. Agree to abandon recent rates of freight and rebates by Railroads.
10. Certificates to be issued for all oil.
11. Transfer of certificate to be a delivery as between Pipe Line & purchaser.
12. Right of action for violation of agreements by any person signing it.

[21] Sherman Papers.

CONTRA.

1. Right to take any rebate which railroad Cos. are at liberty to grant other shippers (See R. R.)
2. Release of all rights of action.

2.

MEMORANDA.

*Feby. 18th.*

*Railroad—Transportation. P. R. R. Contract.*

1. Make known to all shippers, all rates of freight. Open rates.
2. Will not pay or allow any rebate drawback or commission upon petroleum shipments.
3. Greater than that allowed to any other shipper of *like quantity.*
4. Any discrimination shall be *reasonable* and shall, upon demand, be communicated to all persons who are now or may be engaged in shipping.
5. No discrimination in the distribution of cars, but will make fair distribution among all shippers having petroleum to ship.[22]

This summary Sherman projected against a list of actions by Hoyt and Palmer that had delayed and prevented adjudication of the Commonwealth Suits, and of Justice Paxson's intervention in Commonwealth *v.* Rockefeller *et al.* The reservation by the Standard Oil of the right to receive and by the Pennsylvania of the right to grant reasonable rate concessions on larger shipments, which the Pennsylvania Constitution permitted, was the great concession that Sherman and Campbell had had to make in order to achieve the negotiated settlement that Hoyt and Palmer had required. The memorandum did not mention Sherman's victory in requiring that the Standard Oil pay the $40,000 expenses of the Commonwealth Suits. This was the customary obligation of the loser in litigation who resorted to a settlement out of court.

During the two-day meeting of the General Council Campbell telegraphed Governor Hoyt asking his consent to formal withdrawal of the extradition requisitions for the Rockefellers and other out-of-state Standard Oil executives indicted in Commonwealth *v.*

[22] *Ibid.*

Rockefeller *et al.* This action, after approval by the General Council, was obviously part of the negotiated settlement.[23]

Beginning on February 18 Campbell presented to the General Council the agreement whereby the conspiracy Suit had been withdrawn together with the injunction and *quo warranto* proceedings against the United Pipe Lines. He announced the Standard Oil's payment of the costs of the litigation, which sugar-coated the pill of disappointment at having fallen short of the multiple objectives of the Commonwealth Suits. He assured the delegates that the causes leading to the prosecution of the Suits and the grievances listed in the bill of complaints that had been the basis of the conspiracy indictment "are promised to be removed, and entirely rectified, where it is admitted that they have existed."

This was not entirely true. The Standard Oil companies and subsidiaries in their contract with Campbell agreed not to oppose the "entire abrogation" by the railroads of "the system of rebates, drawbacks and secret rates of freight in the transportation of petroleum," or the publication of rates, and they promised not to receive any rebate or drawback "that the railroad companies are not at liberty to give to other shippers of petroleum." The agreement left the Standard Oil free to accept lower open rates based upon its larger shipments. It was this stipulation that Sherman had fought vainly to eliminate from the agreement and to substitute in its stead equal rates on all oil shipments of 1,000 barrels or more. Thus, the General Council had not won its objective of securing equal petroleum rates on the railroads. That section of the agreement left the Standard Oil companies and subsidiaries free to seek new rate concessions based upon their larger shipments and to continue to receive secret rebates if the railroads did not abolish the rebate system.

The real gains that Campbell reported were in pipeline practices, a field in which the Standard Oil no longer held a monopoly because of the successful Tide Water. Among these gains were

[23] *Ibid.*, Lewis Cassidy to Campbell (twice) February 18, 1880, Sherman to Cassidy, February 19, 1880, Campbell to H. M. Hoyt, February 19, 1880; *History of the General Council*, pp. 40–41.

nondiscriminatory acceptance of and piping and storage of all oil at equitable, reasonable rates up to 65,000 barrels maximum for not more than fifteen consecutive days, plus the permanent abolition of "immediate shipment" discount prices. Campbell displayed the contract with the appended signatures of the executives of the Standard Oil group of companies led by John D. Rockefeller.[24]

Campbell's report precipitated an "interesting" debate during the most "lively" meeting yet held in Producers Hall, the *Titusville Morning Herald* reported. Vigorous objections were raised that there was no practical method whereby the Standard Oil companies and subsidiaries could be compelled to fulfill the contract and prevented from reverting to "old customs complained of." Actually the delegates were presented with a *fait accompli*, since the key Commonwealth Suits had already been withdrawn. A majority of delegates asserted that Campbell could not have acted otherwise than he did in negotiating the settlement. The *Morning Herald* praised the disinterested giving of his time and money in the service of the Producers Unions, declaring that no other man would have worked "in this cause so long and with so small encouragement." It then described the decline of the Producers Unions. The agreement with the oil combination, the editor added, was consistent entirely with "the principles" that Campbell had "so long and so consistently advocated." [25]

On the second day the General Council endorsed his discontinuance of the Suits against the Standard Oil combination and the Pennsylvania Railroad "as wise and judicious, and compelled under the existing conditions of the litigations." However, the majority disapproved of the reservation in the contract reserving to the Standard Oil the right to accept discriminatory rates that the railroads were at liberty to give other shippers. The Council thanked Campbell "for the energy, fidelity and ability with which he conducted the several legal proceedings." Although the delegates wished to exonerate him from blame and approved of his leadership

[24] *History of the General Council,* pp. 39–44; *Titusville Morning Herald,* February 19, 1880.

[25] *Titusville Morning Herald,* February 19, 1880.

there was a powerful current of dissatisfaction with the settlement.

After Campbell's report, Roger Sherman, as Chairman of the Resolutions Committee, presented a stinging resolution that excoriated the state administration's "failure to fulfill its promises of support" and attributed this to "the dangerous influence of corporations with the legislative, executive, and judicial departments of the government." This expressed the attitude of Sherman and the delegates' by blaming Governor Hoyt for the General Council's inability to secure adjudication of "the correctness of our position," which the resolution attributed to delays resulting from interference by high state officials, to Governor Hoyt's refusal to approve the extradition requisitions, and to the "extraordinary" interference "of some of the Judges of the Supreme Court" with the trial of Commonwealth *v.* Rockefeller *et al.* In all this Sherman and the General Council, which approved the resolution, perceived "the alarming and most dangerous influence of powerful corporations." While accepting "the inevitable results forced upon us by these influences we aver that the contest is not over and our object not attained, but we will continue to advocate and maintain the subordination of all corporations to the laws, the Constitution and the will of the people." The resolution asserted also that freight-rate discrimination by common carriers was wrong in principle and fostered "dangerous usurpation" and that it was the duty of all branches of government to protect the public "from this growing and dangerous power." [26]

After approving this defiant reassertion of the principles of the producers' movement with but five dissenting voices, a number of delegates "emphatically declined to sign" the contracts with the Standard Oil and Pennsylvania Railroad. Their action left the negotiated settlement in force, however, since Campbell had signed it in behalf of the General Council. Obviously, it left unremedied the problem of excessive production that had attracted such attention during 1878 and 1879, while the long-sought stable market for crude oil was unachieved.[27]

[26] *Ibid.*, February 20, 1880; *History of the General Council,* pp. 46–47.
[27] *Titusville Morning Herald,* February 20, 1880.

That afternoon Sherman reported to Shiras that the General Council had "declined to approve of the terms of the contracts, kicked the state administration and thoroughly approved and vindicated our friend Mr. Campbell." From the $40,000 expense money paid the General Council there would be left for distribution among the Commonwealth Associate Counsel from $20,000 to $22,000 for payment of their fees after paying all bills and the counsel in Clarion and Philadelphia and "reimbursing all contributors." Only some $14,000 had to be refunded to those who had financed the Suits, Sherman had informed a friend. That inadequate sum contrasted vividly with the vast resources of the oil combination and its trunkline allies. Campbell had received merely his expenses and the refund of his contribution. Sherman proposed to Shiras that although he himself needed as a "poor" man "all the money this or any other business yields me, . . . we so fix our fees as to consume nearly this balance, and that we then write a letter to Mr. Campbell, enclosing him a draft for such a sum as we may agree upon, and presenting him with it, with our compliments, as a recognition of his professional services in the cause, and our appreciation of his manly conduct throughout this thankless struggle. Please see Judge Acheson and let me know your views." [28]

The Associate Counsel, in other words, made a present of approximately $6,000 as a fee for legal services to President Campbell from the funds available for their fees, a delicate compliment that attested to the caliber of his collaboration with them in the Suits and his devotion to the principles of law that Sherman, Shiras, and Acheson upheld. When Ohlen attempted to compensate Campbell, Sherman advised him not to accept the gift.[29] The fees of the Associate Counsel were proportionately reduced. Sherman's was the largest, a modest $5,000 compensation for the time and skill that he had devoted during nearly two years to the conduct of the Suits.

On Campbell's initiative the General Council established a Committee on History to prepare a history of itself, its legislative policy,

[28] Sherman Papers, Sherman to Shiras, February 20, 1880.

[29] *Ibid.,* Sherman to Campbell, May 18, 1880, Ohlen to Sherman, April 28, 1880, Sherman to N. C. Clark, March 8, 1880.

and the litigation. Campbell appointed Sherman as chairman. Sherman then asked him for all his "papers, telegraphic and other correspondence of value," and a "statement of all the facts relative." "Of course," he added, "it is absolutely necessary that these exact *facts* should be given correctly, and they should be used prudently and . . . with moderation." This was his reaction to Campbell's insistence that the history "show up the true inwardness of those suits."[30]

Sherman was proud that the Committee on Legal Remedies had paid no newspapers to publish data favorable to the General Council and the Commonwealth Suits. This was an oblique commentary upon the Standard Oil's subsidizing of a large portion of the Oil Region press.[31]

It was not until April 27, after sporadic negotiations, that an agreement was signed by Thomas A. Scott for the Pennsylvania Railroad Company and by Campbell for the General Council of the Producers Unions. In this the Pennsylvania Railroad promised publication of rates on petroleum and not to pay or allow thereafter to any shipper of such products "any rebate, drawback or commission" upon shipments of the same quantity, and that any discrimination favoring large shippers "shall be reasonable, and . . . upon demand . . . be communicated to all persons shipping, or who are now or may be hereafter engaged in the business." That carrier also promised to apportion its oil tank cars fairly among shippers.[32] Similar contracts with the Erie and New York Central systems were not forthcoming, though Sherman to the very last had attempted their negotiation. However, the agreement with the Pennsylvania Railroad secured to independent producers, oil buyers, and refiners in the Keystone State as fair a system of petroleum rates and service as existing circumstances permitted. Like the contract with the Standard Oil group of companies its enforcement depended largely upon the good faith of the independents' former antagonist.

[30] *Ibid.,* Sherman to Campbell, March 22, 1880, Clark to Sherman, February 25, 1880.

[31] *Ibid.,* Sherman to James P. Barr & Co., August 29, 1879.

[32] *History of the General Council,* pp. 38–39, 45–46.

Elisha G. Patterson appeared at the General Council's May meeting as an active member "in the character of Mephistopheles," Sherman reported to Campbell. Patterson offered ten resolutions calling for reports from the Committee on History, Secretary Byrom, and the Committee on Legal Remedies, and a report on the disposition of the expense money paid by the Standard Oil. Knowing that his former friend had stirred up the noncontributing producers, Sherman disliked the insinuations implied by his resolutions. He attacked Patterson at the meeting and told him that if he had any personal curiosity regarding the conduct of the Commonwealth Suits he would tell him gladly all that he knew. There was "nothing to conceal" and "all the facts had been made public." Sherman believed that Patterson's hostility to the Tide Water had motivated his attack upon himself and Campbell, with whom he had broken when the conspiracy suit was launched against Rockefeller. As for Patterson, Sherman informed Campbell: "Unfortunately he lives in a glass house, and if he is not careful two persons to whom he confided a by-no-means honorable scheme of his may be obliged in self defense to tell what it was." When, he added, were the Campbells going to visit the Shermans as they had promised to do? [33]

Campbell was discouraged, "even bluer than ever," after learning how much neglect of his producing business had cost him during the period of his leadership of the General Council. That included the alcoholism of his partner in a refinery that had produced financial loss. Campbell replied to Sherman:

> I am not surprised at Patterson's course—he considers the whole attack on the Standard as a magnificent chance of extorting money. He was mad when I disclosed Jones testimony as he said that gave the case away and the action for conspiracy took away all chance of profit by privately threatening & compromising & abandoning the cause. I also fully appreciate his bravery in making the attack in my absence.

Campbell asked that the Committee on History be ready to report at the General Council's next meeting, "a mere statement of facts showing how we have tried hard & fast but be careful not to

[33] Sherman Papers, Sherman to Campbell, May 18, 1880.

compromise yourself with the Supreme Court." He would himself "show to any enquiring minds" how the $40,000 expense money had been distributed. "Please keep me posted & if the parties you refer to should show our friend up in his true colors I will not cry." [34] As Campbell drafted his "recollections of the fight" for the Committee on History he enjoined Sherman: "Keep your eye on Pat 'Did he catch the cat.' Memo: The above was not meant for poetry." [35]

In an interview after the mid-June General Council meeting Patterson expressed his bitterness to Sherman at not having been paid for his year of work for the Producers Unions in behalf of their legislative program and their litigation. He offered his draft version of the movement's history "up to April 23d 1878," which Sherman gladly promised to read, while assuring him that the forthcoming *History* would not criticize him for withdrawing from the Committee on Legal Remedies. "I think his position is utterly absurd and untenable, that he has no backers, and the best way is to ignore and freeze him out," Sherman wrote Campbell. "Patterson," the latter replied, "is given to lying." By that time Sherman's "report" from the Committee on History had been approved by the General Council and 1,000 copies ordered printed. "You had better explode that torpedo you have planted under Patterson," Campbell added.[36]

Meanwhile, Governor Hoyt had nominated George A. Jenks for the Pennsylvania Supreme Court with the backing of Sherman and his friends. That was a reward for his willingness to assist the prosecution of the Standard Oil executives in Commonwealth *v.* Rockefeller *et al.*[37]

Sherman's *History of the Organization, Purposes and Transactions of the General Council of the Petroleum Producers' Unions, and the Suits and Prosecutions Instituted by it from 1878 to 1880* was published in Titusville that summer. Written in the inimitable

[34] *Ibid.,* Campbell to Sherman, May 21, 1880.

[35] *Ibid.,* Campbell to Sherman, June 1880.

[36] *Ibid.,* Sherman to Campbell, June 16, 1880, Campbell to Sherman, August 16, 1880.

[37] *Ibid.,* Sherman to G. A. Jenks, May 3, 1880.

military style of an ex-Confederate cavalryman, it was, as A. N. Perrin, a fellow committee member and prominent producer, observed, "a fair history of the case although it hits pretty hard now and then & possibly may hurt." [38] Distributed by Secretary Byrom, it became the official apologia of the Producers Unions. Restrained, compressed into a pamphlet, devoid of personalities, it omitted modestly Sherman's central role in the origin and conduct of the Commonwealth Suits.[39]

After summing up the "concessions" wrested by the Producers Unions from the Standard Oil combination, and conceding that reasonable rebates on large petroleum shipments would probably be upheld in the courts, Sherman admitted that but few individual producers had signed the agreements with the Standard Oil and the Pennsylvania Railroad with a view to availing themselves of their provisions. What, he asked, had the Producers Unions and the General Council accomplished in their long campaign?

First of all they had exposed the system of legislation and government that had produced great corporations built upon "discrimination and favoritism" destructive of competition while undermining "the foundations of free government" by destroying popular liberties. They had revealed the methods that had produced the Standard Oil monopoly, and had persuaded "the public mind . . . to think upon the means of remedy, and to believe that such combinations of corporate capital must be controlled by stringent and easily administered law." They had extorted from "the wrong-doers an implied confession of the unlawfulness of their acts, and promises of future good behavior," while the great railroads had been converted to the belief "that the rebate system is as bad for them as for the public." Unless "the signs of the times fail to be prophetical, in the near future the United States will exercise its constitutional right to regulate commerce between the States, and in the States themselves laws will be enacted, and such men be elected to enforce them, as

[38] *Ibid.*, Perrin to Sherman, August 12, 1880.

[39] Sherman sent complimentary copies to friends on September 8. For Sherman's admission of authorship, cf. Sherman Papers, Sherman to W. C. Clark, September 8, 1880, Sherman to H. D. Lloyd, July 15, 1880.

will bring the great corporations to obedience to the principles of justice and true public policy." [40]

While this eloquent peroration transformed Sherman's *History* into a tract advocating the railroad regulation that the independent producers demanded, Governor Hoyt reached a similar conclusion. In his biennial message to the legislature of January 4, 1881, he reported that despite the belated Pennsylvania Railroad adhesion to the state Constitution's requirements of common carriers while "prosecuting their business with oil producers," statutory protection was needed for shippers so that the courts could enforce the Constitutional provision upon "all transportation companies, at all times in all places," and in behalf of all interests.[41] Hoyt was loyal to the cause of an impartial transportation policy after all. A year earlier before the congressional Committee of Commerce, Franklin B. Gowen had developed this implication of the Commonwealth Suits, when he demonstrated that judicial action alone was an inadequate remedy for railroad malpractices.[42]

The final settlement left many producers bitter. They were convinced that they had been betrayed by the state administration, an opinion that Sherman's *History* confirmed. Some angry Bradford producers accused Campbell of having "sold out to the Standard Oil for fabulous sums." That made him so angry that he stopped fighting that business giant. Thereafter he directed the Bear Creek Refinery at Pittsburgh, which he had organized as part of the independent movement, happy that he could ship his oil "in almost any direction, just as cheap as the Standard Oil Company." This indicated, he asserted, that the independents had won more than "a barren victory." His refinery, he added, would be "good business if it was only profitable. However, it is far more profitable than fighting the battles of producers and getting curses instead of thanks for your work in their behalf." As late as May 1887 Sherman found it necessary to defend him and the General Council from similar accusations. Later that year, a committee of his own friends headed

[40] *History of the General Council,* pp. 49–50.
[41] *Papers of Governors,* IX, 773–775.
[42] Gowen, *op. cit.,* p. 24.

by David McKelvy published in the *American Citizen* an open letter vindicating Sherman too from aspersions on his conduct of the negotiation that had terminated the Commonwealth Suits.[43]

While Sherman and the Producers Unions with Patterson's help had initiated the movement that produced the Interstate Commerce Act of that spring, he had been much discouraged late in 1880 at the people's inability to "help themselves to get from under any weight." He retained then his contempt and antipathy for the Standard Oil. Despite his own discouragement, and admitting that he felt "intimidated," he backed quietly the attempt of Campbell and northwestern Republican producers to elect George Shiras, Jr., United States Senator. Although western Pennsylvania backed him strongly it was unable then as later to overthrow Senator Don Cameron's control of the Republican party or to defeat the dominant Philadelphia and central Pennsylvania interests which he represented.[44] Shiras continued for twelve more years to be the leading member of the Pittsburgh bar, a model of professional integrity and proficiency.

[43] *Petroleum Age,* I (April 1882), 165; *American Citizen* (Titusville), October 7, 1887; Sherman Papers, Campbell to Sherman, May 23, 1887.

[44] Sherman Papers, Sherman to Campbell, December 23, 1880.

# CHAPTER IX

# *Completion of the Settlement*

THE contracts with the Standard Oil and the Pennsylvania Railroad which President Campbell signed had made competition with the former's group of companies theoretically possible in oil buying, shipping, refining, and marketing of petroleum products. Successful Tide Water-Reading competition had deprived the Rockefeller combination of its complete monopoly of trunkline oil transportation to the seaboard. Yet, the Standard Oil's subsequent pool with its new rival left it in control of approximately 80% of that traffic. Furthermore, the Standard Oil was constructing rapidly long-distance crude-oil pipelines radiating from the Oil Region that within two years would deprive the railroads of most of their crude-oil traffic. The Standard Oil's significant share of oil production, its system of gathering pipelines, its control of the tank cars of the two northern trunklines, its 80% of the nation's refining capacity, and its nationwide marketing organization stood unshattered by the attacks of the independent oil men. Impartial administration of its pipeline system was enforceable at law only in behalf of those independents who had signed the contract with it. Those were few. The Pennsylvania Railroad's contract with Campbell was as enforceable as were the short-lived trunkline pooling agreements. Scott would break it if the northern trunklines continued their favoritism to Rockefeller.

Actually, Campbell and Sherman had won a moral victory against pathological business methods in the petroleum industry and on the railroads instead of decisive abolition of those evils. After two and a half decades more of experience with railroad discrimination and semipiratical industrial management the Progressives of 1903 to 1917 would campaign likewise for effective railroad regulation and restoration of competition by reform of entrepreneurial practices.

The campaign of the independent oil men from 1877 to 1880 had had four aspects. Of these, the fortunes of their seaboard pipelines, the producers' legislative program, and the Commonwealth Suits have been described. The fourth aspect, the attempt of independents to wrest proportionate damages from the Pennsylvania Railroad Company because of illicit favoritism to the Standard Oil was of almost equal importance. Victory in such litigation under the Pennsylvania Constitution of 1873 might make rate and service discriminations too costly for continuance by the trunklines. As in the legislative program and the Commonwealth Suits, the solution sought was essentially a lawyer's remedy. In the attempt to achieve this fourth objective Sherman again played a major role.

There remained also the question of the disposition of the gathering pipeline system of the bankrupt Pennsylvania Transportation Company, which was managed by a receiver when the Commonwealth Suits were launched. If this system was actually independent, pipeline competition in the Oil Region would be strengthened. If the system were acquired by the Standard Oil, competition would be severely impaired and the position of the Tide Water jeopardized. In the litigation between the creditors and receiver of the Pennsylvania Transportation Company, but not in the final disposition of its assets, Sherman played an important part. This brought him again into close relationship with the Philadelphia capitalists who had helped to finance Henry Harley's great venture.

When these matters were settled there remained the question of what, from the viewpoint of John D. Rockefeller, should be done about Roger Sherman. Informed persons knew that he had been the inspirer of the independent movement. His astute legal brain had exposed the secret methods whereby monopoly had been substituted for free enterprise in the petroleum industry and the railroad oil traffic, and he had also helped to plan the independent pipelines' and the producers' legislative program.

In the direct attack by means of damage suits upon railroad discrimination in the oil traffic Henry C. Ohlen played the initiating role. He was the leading independent oil dealer of New York City. He shipped crude oil from the Oil Region to New York for such independent refiners there as Lombard & Ayres and Jules Rousseaux, as well as for export. Ohlen had been in the business since 1865. In 1872 he led the New York opposition to the South Improvement Company in a manner that exposed Jay Gould's railroad favoritism in the oil traffic. He contributed moderately to the legal expense fund of the General Council of the Producers Unions. He presented important testimony before General Sweitzer in Commonwealth *v.* Pennsylvania Railroad.

Before the producers' appeal to Governor Hartranft Ohlen had petitioned in the Supreme Court of New York in New York City for a writ of mandamus to compel the New York, Lake Erie & Western Railroad to supply him with tank cars for transportation of his crude oil. Ohlen alleged that the defendant had declined to supply him with tank cars when they stood idle because it had entered into a combination with Charles Pratt & Company, a Standard Oil subsidiary, to confer upon it exclusive use of that carrier's tank cars. Ohlen's petition was rejected because of a technicality despite the skill of his counsel, Evarts, Southmayd & Choate.[1]

It was a compliment to Roger Sherman, therefore, that Ohlen retained him, after he had presented the *Appeal to the Chief Executive of Pennsylvania* to Governor Hartranft, as counsel in behalf of himself and the "outside refiners" of the New York area in a suit for damages against the Pennsylvania Railroad based upon its covert rebates to the Standard Oil. Elisha G. Patterson had advised Ohlen to secure fulfillment of his contract with the Empire Transportation Company that the Pennsylvania Railroad had assumed, after the latter rejected an application for a rate refund proportionate to the rebates granted the Standard Oil. Ohlen was determined to test the matter by invoking the arbitration clause of his contract, and was convinced that his claim would open up evidence on the

[1] Sherman Papers, Ohlen to Sherman, January 4, October 9, 1878, MS petition, July 9, 1878, MS reply of the New York, Lake Erie, and Western Railroad Company, Ohlen to National Storage Company, December 2, 1878.

basis of which he could sue for a "much larger claim for discriminations." He retained Patterson as his referee and Sherman as his counsel for the arbitration. After discovering that Ohlen's contract did not authorize compulsory process Sherman advised him that a bill in equity after the damage suit was brought would "compel a discovery of the Evidence. . . . There would be no escape from this and by interrogatories you could probe the matter to the bottom." [2]

Proceedings were begun first against the Empire Transportation Company, then in liquidation. Although Ohlen supplied Patterson and Sherman with his complete correspondence with Joseph D. Potts, President of the Empire, and a list of cars and amounts of oil on which he claimed rebate and the refund, the proceeding was handicapped by his inability to specify the exact form in which the Standard Oil had received a rebate equal to the pipage charge from the Empire. Ohlen agreed with Sherman that the best course was to pursue arbitration until "no relief can be had by that means" and then go to court. Jules Rousseaux, who retained Sherman for a parallel damage suit against the Pennsylvania Railroad, supplied evidence of having been charged full rates by it.[3]

Sherman explained to Rousseaux the legal basis for a suit to recover freight rate payments "in excess of that charged the Standard Oil Co." Common carrier rates must be "*equal* to all shippers" and "lawful" discriminations must be "reasonable, and based upon direct advantage to the carrier." A common carrier could not employ its corporate powers "to build up one shipper or class of shippers to the injury of another." Since the lowest rate charged on a specific type of freight presumptively was "a reasonable compensation to the carrier" a higher rate was "unreasonable," and if paid

[2] *Ibid.,* Ohlen to Sherman, September 26, 1878, April 1, 1879, Sherman to Ohlen, September 24, 1878, Ohlen to Patterson, September 17, 1878; Commonwealth *v.* Pennsylvania Railroad Company, *Printed Testimony,* p. 579.

[3] Sherman Papers, Sherman to Ohlen, September 24, 1878, Ohlen to Sherman, September 26, October 7, 26, December 12, 1878, Ohlen to Patterson, October 7, 1878, J. Rousseaux to Sherman, December 11, 13, 1878, W. G. Gray to Rousseaux, December 11, 1878.

"under protest" the protesting shipper could recover the amount "so paid in excess of a reasonable rate." Rousseaux's claim for such refund should be from $50,000 to $80,000, Sherman advised. He should bring suit immediately for this in the United States Circuit Court. "It does no good to negotiate," Sherman said as he requested an itemized statement of quantity and carloads shipped and from what points, and a retainer fee of $500.[4]

For Ohlen, Sherman first attempted, vainly, to negotiate a settlement of his claim with the Pennsylvania Railroad. Receiving no reply to his overtures, he advised Ohlen and his associates to bring suit in the United States Circuit Court, while continuing on current shipments to reserve the right to claim the same rebates as were allowed "to any other shipper." To this they agreed, and sent him a $500 retainer.[5] Sherman then associated Mark Acheson and George Shiras, Jr., with himself as counsel in a $200,000 damage suit commenced in that court at Pittsburgh in Ohlen's behalf. This action was based upon the theory of recoverability of excess charges. In this and the Rousseaux suit Sherman insisted that his clients present specific evidence of having paid higher rates under protest.[6]

Sherman left General Sweitzer's hearings in the Commonwealth Suits hurriedly on December 11, 1878, with Shiras to confer with the Ohlen associates in New York City about this litigation and to advise on the rehearing of Ohlen's petition for a mandamus against the Erie railroad. This produced Sherman's first appearance before the New York Supreme Court and brought him into intimate relationship with the New York independents.[7]

The basis of Ohlen's suit against the Pennsylvania Railroad was the common law rights of shippers. Although that carrier was understandably suspicious as to the relation between Ohlen *et al. v.*

[4] *Ibid.*, Sherman to Rousseaux, December 16, 1878.

[5] *Ibid.*, Ohlen to Sherman, November 23, 26 (twice), December 14, 1878, Sherman to Ohlen, December 2, 1878, Sherman to Pennsylvania Railroad Company, November 23, 1878.

[6] *Ibid.*, Sherman to Acheson, December 20, 23, 1878, Sherman to Ohlen, December 23, 1878, Ohlen to Sherman, January 4, 1878, Acheson to Sherman, December 23, 26, 1878.

[7] *Ibid.*, Ohlen to Sherman, December 11, 1878, July 5, 7, 1879.

Pennsylvania Railroad Company and Commonwealth *v.* Pennsylvania Railroad Company, Hampton's quizzing of Campbell before General Sweitzer revealed that Sherman, Acheson, and Shiras were acting independently in the former. Ohlen promised Sherman liberal compensation if his suit succeeded, but desired that the expense "be as light as possible." [8]

In the parallel Commonwealth Suits and Ohlen *et al. v.* Pennsylvania Railroad Company Sherman and his associates employed evidence gathered in either to the advantage of the prosecution or plaintiffs in the other. The disclosures of specific rebates paid by the Pennsylvania Railroad that were made in the testimony presented to General Sweitzer were of material assistance to Sherman in drafting the advance "declaration" of the Ohlen suit for the May term of the Circuit Court. When Sherman learned in preparation for trial of Rousseaux *v.* Pennsylvania Railroad Company of the latter's denial of tank cars to the former, Sherman put Rousseaux's oil buyers on the witness stand before General Sweitzer in Commonwealth *v.* Pennsylvania Railroad Company. Sherman also advised Ohlen to institute a second suit against that railroad to recover overcharges paid under protest since the first suit had been filed. "In any negotiation looking to a cessation of hostilities," Sherman added, "you should insist upon being made good for the large sums of money of which you have been robbed, and the evidences of which robbery is now within your reach." This evidence made Ohlen hopeful of being able to wrest large damages from the Pennsylvania Railroad by jury trials.[9]

When the Ohlen associates consulted with the Producers' Committee that interviewed the Standard Oil, Ohlen advised Sherman that no contemplated settlement of the Commonwealth Suits would affect his suit against the Pennsylvania Railroad, nor would he make any settlement with it or the Standard Oil that would "involve the

[8] *Ibid.,* Ohlen to Sherman, January 4, 1879.

[9] *Ibid.,* Sherman to Ohlen, January 9, March 24, 1879, Ohlen to Sherman, March 26, 1879, W. G. Gray to Sherman, January 20, 1879; Commonwealth *v.* Pennsylvania Railroad Company, *Printed Testimony,* pp. 65–77.

loss of the money which I believe to be justly due." Although his object was to purchase oil as cheaply as he could get it, he said he did not wish to assume "a hostile position towards the railroads, as I need them for purposes of transportation. . . . I only insist that they shall treat me in the same manner, and give me equal privileges with every body else." The outside refiners, also, would not consent "to the consolidation of the oil interests." [10]

Sickness in the Sherman family from January to March 1879 retarded action in the case of Ohlen *v.* Pennsylvania Railroad Company as it did the Commonwealth Suits. Sherman informed Ohlen that the indictments found in Commonwealth *v.* Rockefeller *et al.* had no bearing upon or connection with his suit. In reply Ohlen instructed him to initiate the second suit against the Pennsylvania Railroad for refund of overcharges "covering December to May, inclusive." This Sherman did in an action for $150,000 damages with a listed overcharge claim for $99,234 for those six months. "I suppose you are not suffering from any rebate *just now,*" Sherman remarked to Ohlen with reference to the trunklines' oil rate war against the Tide Water-Reading, "and hope you are in a position to avail yourself of the present *low open* rate." He and Ohlen met frequently in the Oil Region to plan for the trial of both suits.[11] Sherman invited Acheson to several days' conference on the same subject at Titusville, and recommended taking testimony under a bill in equity with a view to trial of both suits at the November term of the Circuit Court.[12]

On July 7 Ohlen sent Sherman a copy of the adverse decision in the second mandamus proceeding against the New York, Lake Erie & Western Railroad Company before the New York Supreme Court. The oil buyer remarked: "You cannot fail to understand that it will be regarded by all the roads involved in the suits in progress as a substantial victory for them. . . . It will I fear have a bad effect as regards the suits you now have in hand against the Penn.

[10] Sherman Papers, Ohlen to Sherman, March 26, 1879.

[11] *Ibid.,* Sherman to Ohlen, March 31, May 5, June 30, 1879, Ohlen to Sherman, April 1, May 1, June 3, 25, 1879.

[12] *Ibid.,* Sherman to Acheson, July 9, 1879.

Railroad, and I am inclined to be still more urgent in the suggestions made to you in our recent interview." He predicted pessimistically that all the Commonwealth Suits "must eventually fall from their own weight" and that "as things now stand no good could come to anybody in the direction sought if they should succeed" because of the existing low schedule of oil freight and pipage rates. Of course, if these should be restored later to a higher level "with rebate the same remedy is open to us as we have at present exercised." Ohlen feared lest a settlement of the Commonwealth Suits leave him alone confronting the Pennsylvania Railroad. "I rely upon you," he wrote Sherman, "that no settlement shall be effected without my claims receiving due consideration." [13]

Ohlen was needlessly discouraged. As Sherman informed Acheson, the New York Supreme Court had quashed the mandamus against the Erie chiefly because Ohlen possessed "an adequate remedy at law by an action for damages." [14] Sherman assured him that his suggestions were "carefully considered" and that his interests would be "attended to in the connection you mention." Acheson observed that the negative result of the mandamus proceeding made it "only the more necessary to push with vigor the pending suits for damages" against the Pennsylvania Railroad. He promised to join Sherman in "all necessary measures" in preparation for a trial in November.[15]

Meanwhile, Sherman requested of Colonel Potts that he propose "prompt settlement of the excess freight" that Ohlen had been obliged to pay under protest in violation of his contract of April 2, 1877, which the Pennsylvania Railroad had assumed. When Potts took refuge in the Empire's incomplete dissolution Sherman reminded him that the rights of its creditors were unaffected, and included him as its President in a third damage suit against the Pennsylvania Railroad Company instituted in the Philadelphia County Court of Common Pleas with the aid of Isaac Meyer, a

[13] *Ibid.,* Ohlen to Sherman, July 7, 1879.

[14] *Ibid.,* Sherman to Acheson, July 10, 1879.

[15] *Ibid.,* Sherman to Ohlen, July 11, 1879, Acheson to Sherman, July 14, 1879.

competent Philadelphia attorney. Supervision of this suit was added to Sherman's heavy schedule. Reviewing the Philadelphia Common Pleas Court's irregularity in handling the Empire's dissolution, he instructed Meyer to secure able associate counsel and make a "strong fight." [16]

When the Standard Oil attempted to initiate negotiations in October 1879 for settlement of the Commonwealth Suits, Sherman informed Ohlen that it might soon become "greatly to the interest of the Penn. R. R. Co. to pay . . . claims for excessive freight charges." Until then, Sherman told him: "It is best that this should be more than ordinarily confidential, and that you wait in readiness for such action as may best subserve your interests." Ohlen was apprehensive lest he be called to testify before the Hepburn Committee to the damage of his suits against the Pennsylvania Railroad, and informed Sherman that in such a contingency he wished Sherman to be his counsel during his "Examination." This expressed great confidence in Sherman's professional competence.[17]

Sherman was ready in November for trial of the first two suits against the Pennsylvania Railroad. Acheson was to take testimony in Philadelphia in the third suit against the railroad with a view to trial in May. Sherman advised Ohlen not to agree to any postponement after certain parties had advised him that his "interests" would "be largely promoted . . . by a further postponement." Actually, evidence uncovered by the Commonwealth Suits and the Hepburn investigation that the Standard Oil had been paid a drawback upon 30,000 barrels of oil that Ohlen had shipped made it certain that he would win his suits in the Circuit Court.[18]

[16] *Ibid.*, Sherman to I. Meyer, August 4, September 15, 1879, Sherman to Ohlen, July 29, August 4, September 8, 15, 1879, Sherman to Potts, July 24, 1879, Ohlen to Sherman, September 6, 1879, *Paper Books,* IX, *Henry C. Ohlen v. The Pennsylvania Railroad Company and J. D. Potts, President, Empire Transportation Company, Court of Common Pleas, County of Philadelphia, In Equity.*

[17] Sherman Papers, Sherman to Ohlen, October 6, 1879, Ohlen to Sherman, October 7, 1879.

[18] *Ibid.*, Ohlen to Sherman, n.d., October 27, 1879, Sherman to Ohlen, October 22, 27, 1879; Nevins, *Study in Power,* I, 312.

Ohlen financed the printing of long excerpts from the testimony taken by General Sweitzer for use as exhibits in his own suits, while reiterating to Sherman his hope "for such a turn of affairs as will cause a settlement all round." The Pennsylvania Railroad was obliged to consent that such evidence could be used in Ohlen's suits and that his counsel would "be furnished with copies of settlements between the Railroads" covering the period embraced by the litigation. These extraordinary concessions were proof of the skill of Ohlen's counsel and of the weakness of the Pennsylvania Railroad's defense. Sherman asked Shiras to select "what we want" from an unbound copy of the testimony in Commonwealth *v.* Pennsylvania Railroad Company and other Commonwealth Suits.[19]

On November 10 he added, revealingly, "It is hard to tell whether we should try or not. Ohlen does not want to. On account of my own personal convenience I do not. I presume the P R R has notified the Standard that it must settle this claim and I think it is understood that it shall do so." If the first Ohlen suit were continued it should be on the Pennsylvania's application "with a distinct understanding that it is to be tried at May term, and we should get ready so far as arguments &c are concerned."[20] At that time Sherman was engaged in preparations for trial of Commonwealth *v.* Rockefeller *et al.* and in the vain attempt to persuade Governor Hoyt to approve the extradition requisitions. Sherman ordered Meyer to use the index to the testimony in Commonwealth *v.* Pennsylvania Railroad Company for data "as to rebates &c" for use in Ohlen *v.* Pennsylvania Railroad Company and J. D. Potts, President, Empire Transportation Company, adding that he would confer with him in the Continental Hotel, Philadelphia, on January 4. Sherman's procedure in all three cases was to establish the facts and the right of Ohlen's claims by filing a careful selection from the printed testimony in Commonwealth *v.* Pennsylvania Railroad Company so that formal trial could be confined to arguments by

[19] Sherman Papers, Acheson to Sherman, November 6, 1879, Shiras to Sherman, November 6, 1879, Sherman to Shiras, November 10, 1879.

[20] *Ibid.,* Sherman to Shiras, November 10, 1879.

opposing counsel of the principles of law and equity upon which the courts' decisions must rest.[21]

Sherman conferred with Ohlen and Meyer in Philadelphia before the Commonwealth Associate Counsel opposed the writ of certiorari in Commonwealth *v.* Rockefeller *et al.* before Justice Paxson in January. Sherman and Ohlen conferred a day later in New York City, when the Titusvillian resumed his secret negotiation with Heydrick for a settlement between the independents and the Standard Oil. Ohlen sent Sherman a copy of the Rockefeller coterie's contract with H. L. Taylor & Company, oil producers. Ohlen had worked during the New York negotiations to produce a successful settlement, offering a "liberal" compromise of his suits with the Empire and Pennsylvania.[22]

Immediately after Sherman and Campbell concluded the contract settlement with the Standard Oil, Ohlen gave Sherman a power of attorney on January 30 "to negotiate for a settlement of all my claims" against the Pennsylvania Railroad and Empire Transportation Company. Three months of negotiations in which Shiras and Acheson also figured produced on April 21 a settlement of "the Ohlen cases for $131,250." Of this sum which was paid to Shiras on Ohlen's account that attorney retained 10% for fees of counsel. Of this Sherman received $5,000, Shiras $4,125, and Acheson $4,000. From his net proceeds Ohlen refunded to B. B. Campbell his contribution to the producers' legal expense fund in recognition of his services during the negotiation. After this Sherman returned to Ohlen all the documents and papers gathered in preparation of the three damage suits.[23]

[21] *Ibid.,* Ohlen to Sherman, November 11, 12, 1879, Sherman to Meyer, December 1, 1879, Sherman to Shiras, December 29, 1879.

[22] *Ibid.,* Ohlen to Sherman, December 30, 1879, January 7, 9, 10, 12, 1880.

[23] *Ibid.,* Ohlen to Sherman, January 30, 1880, Shiras to Sherman, April 21, 1880, Sherman to Shiras, February 2, 1880, Sherman to Acheson, February 2, 1880, Sherman to Ohlen, April 24, July 14, 1880, Ohlen to Sherman, April 28, May 8, 1880. Shiras did not think that Ohlen and his associates were entitled to additional damages from the Empire. The three Ohlen suits had been for a total of $350,000 damages.

Although payment to the Ohlen associates in refund of excess charges was modest it was sufficient to establish the principle that railroads were liable to shippers that they discriminated against. This implemented at long last the common law obligation of common carriers that they must accord shippers equitable treatment. In its practical implications this victory in the Ohlen litigation had the most far-reaching significance of all the litigation undertaken by the independent oil men under Sherman's leadership during the years from 1878 to 1880.

The litigation produced by the liquidation of the Pennsylvania Transportation Company was also brought in equity. As an expert in that field Sherman was drawn into it as well. Participation in the litigation enhanced his understanding of the challenge that amorality presented to the business world. He learned also that the disposition of that company's pipelines in the Upper and Lower Districts of the Oil Region would affect the competitive pattern of the petroleum industry.

The Pennsylvania Transportation Company had been in a friendly receivership since October 1876 when M. W. Quick was appointed receiver on Henry Harley's application. Previous to institution of the receivership Harley and Abbott had both borrowed from their company, possibly to protect or finance oil speculation.[24]

From the day of the Pennsylvania Transportation Company's suspension Sherman was counsel for different groups of claimants against it, devoting much time to this difficult, prolonged work until the final settlement of the company's affairs in 1881. Three main groups of creditors battled for five years over the shrunken assets. Among these assets were the partially idle pipes, the oil in storage, and sums due on defaulted contracts for pipage with the railroads. Harley's intention to regain control of his pipeline empire, after

[24] *Ibid., Paper Books,* VIII, No. 2, pp. 23–24, Appendix, pp. 30–31, *Robert W. Mackay v. Pennsylvania Transportation Company, in Appeal of James E. Brown in George H. Yarrow et al. v. Pennsylvania Transportation Company et al.; ibid.,* IX, No. 18, pp. 30–38, *Appeal of James E. Brown in Frederick W. Ames, Trustee v. Pennsylvania Transportation Company et al.,* states that the certificate over issue at the close of 1875 was for but 29,522 barrels.

shaking out the junior creditors, was indicated by the friendly receivership.[25]

Complicated litigation ensued. Quick, the receiver, attempted to collect overdue pipage from the railroads. Three groups of mortgage bondholders and the owners of unredeemed Pennsylvania Transportation Company oil certificates sought to establish the priority of their claims against the day of liquidation. In November 1877 the sale of the Empire Transportation Company's Union Pipeline to the United Pipe Lines Company left only a single buyer for the extensive gathering lines of the Harley system, if it should be decided to sell them rather than reorganize and continue the company as a competitor of the Standard Oil's near-monopoly of the business. As the wells served by the Harley system declined in production and the United Pipe Lines laid new lines to newly developed districts, the longer either step was deferred the greater became the depreciation of the company's competitive strength and the value of its pipelines.

Sherman was immediately retained by certain holders of the prior $200,000 Pennsylvania Transportation Company first-mortgage bonds. These were Charles Barstow Wright, President of the Northern Pacific Railroad Company,[26] and Kemble, President of the Philadelphia People's Bank, in Frederick W. Ames, Trustee, *v.* Pennsylvania Transportation Company *et al.* and in Robert Mackey, Trustee, *v.* Pennsylvania Transportation Company. During litigation between the trustees of the second- and third-mortgage bond issues it became evident that prominent company executives had embezzled money during a manipulation of the securities.[27]

[25] *Ibid., Paper Books,* VIII, No. 1a, *Henry Harley and William Warmcastle v. Pennsylvania Transportation Company,* Appendix, pp. 25–27.

[26] Thomas C. Cochran, *Railroad Leaders 1845–1890* (Cambridge, 1953), p. 500.

[27] Sherman Papers, Sherman to S. G. Thompson, June 18, 1879, *Paper Books,* IX, No. 18, pp. 36–40, *Appeal of James E. Brown from Frederick W. Ames, Trustee v. Pennsylvania Transportation Company et al.,* Appendix, p. 32; *ibid.,* VIII, No. 2, "Testimony in behalf of Defendant," *Appeal of James E. Brown in Robert W. Mackay, Trustee v. Pennsylvania Transportation Company,* Appendix, pp. 23–24, 30–31.

At first, as counsel of Kemble, Wright, and Mackey, Trustee, Sherman vainly attempted to secure the appointment of "a reliable & responsible man" as receiver, while preventing an early sheriff's sale of the company's assets under some indefensible executions. Sherman was obliged to inform his Philadelphia clients that Harley had "fixed" the receivership, which he termed farcical. Prior to the company's failure, its oil certificates had been offered in the market for $1.10 per barrel when the market price of oil was $3.07, indicating a 200% excess issue above the oil actually in storage. As counsel for Wright and Kemble Sherman carried the fight to the Pennsylvania Supreme Court in a vain attempt to supplant Quick as receiver with R. H. Derrickson, a reliable man.[28]

In November 1878 Sherman was counsel in a group of equity cases against the Pennsylvania Transportation Company, for which testimony was taken before a master. From this Sherman learned about secret pipeline-railroad relationships, about the pipage (or commissions) that railroads paid pipeline companies delivering oil to them for shipment, and more about the secret liaisons of the Harley and Abbott combination with the Erie and Pennsylvania railroads. This was invaluable for his conduct of the Commonwealth Suits. Sherman succeeded in preventing Harley from regaining control or from salvaging anything from the wreckage of his bankrupt pipeline empire.[29]

The technical difficulties presented by this equity litigation taxed Sherman's expert knowledge. His attempts to salvage the company's assets for the leading groups of creditors disclosed a high conception of business ethics. The Erie Railroad's bankruptcy and reorganization prevented recovery by the receiver of $359,371 pipage and accrued interest owed by it to the Pennsylvania Transportation Company. Of the $197,285 pipage and interest owed it by the Oil Creek and Allegheny River Railroad, only $58,000 was recovered by litigation. In certain creditors' suits in which he was not an ini-

[28] Sherman Papers, Sherman to Francis Fitz, October 30, 1876, Sherman to W. H. Kemble, November 9, 11, 1876, Sherman to R. W. Mackay, October 25, 1876.

[29] *Ibid.,* Sherman to S. G. Thompson, November 4, 1878.

tial counsel Sherman was invited to argue moot legal points on appeal. The burden of the Commonwealth Suits did not prevent him from continuing to represent Wright and Kemble in successive legal contests for control of the company's assets. His fee in an action before the Supreme Court, in which he defeated a lesser claimant, was but $750. He was not plundering the creditors by exorbitant fees. He had opposed this practice eighteen months earlier when representing Quick, the receiver, in excepting successfully to a master's report authorizing exorbitant fees for counsel of competing claimants that would have absorbed more than half the $58,000 recovered from the Oil Creek railroad. On that occasion Sherman derived from the staff of the Pennsylvania Transportation Company valuable data on Erie and Atlantic and Great Western railway policies that he used in the Commonwealth Suits.[30]

A month later he attempted to have Quick removed as receiver because of his lease of the Lower Division's lines to the United Pipe Lines "without any order of court." To Wright he reported; "The *precise consideration* of this lease is not known . . . and the whole thing is somewhat mysterious." In April 1879 Sherman informed another attorney, J. B. Brawley of Meadville, that the "latest P. T. steal" had been "arranged by agreement in open court at the solemn hour of 11 P.M." before Judge Charles E. Taylor. "What jurisdiction has Judge Taylor in the matter sitting at Franklin? *Possibly* by consent, but not otherwise. I will be on hand," he promised, to assist in reopening and nullifying that deal.

A year earlier in Ames, Trustee, *v.* Pennsylvania Transportation Company he had established in the Crawford County Court of Common Pleas that that company's first-mortgage bonds were a first lien upon its Upper Division pipelines and a second lien upon those

[30] *Ibid.*, Sherman to Brown, January 1, 1879, Sherman to W. R. Bole, February 17, 1879, Sherman to C. E. Taylor, February 20, 1879, *Paper Books,* IX, No. 18, p. 64, "Testimony in behalf of the Defendant," *Ames, Trustee v. Pennsylvania Transportation Company et al. in Appeal of Pennsylvania Transportation Company, Appeal of James E. Brown,* Appendix, *passim; ibid.*, VIII, No. 2, *Appeal of James E. Brown in George E. Yarrow et al. v. Pennsylvania Transportation Company et al., passim; Titusville Morning Herald,* April 29, 1876.

of the Lower Division.[31] This he capitalized upon in the late spring of 1879 in a complicated, skillful negotiation. Wright and Kemble were attemping to secure the appointment of a new receiver of the Pennsylvania Transportation Company by the Philadelphia County Court of Common Pleas, since their position was improved by the discovery that G. U. Anderson, a director of the company, had obtained $19,000 of first-mortgage bonds fraudulently prior to Quick's receivership. Since Quick would object to the jurisdiction of the Philadelphia court and recognized but $125,700 of the first- and second-mortgage bonds as valid company debts, Sherman secured from Wright and Kemble authorization to negotiate a settlement covering all legal points at issue. He conferred with Quick and his counsel, while blocking the receiver's attempt to take the $58,000 recovered from the Oil Creek railroad from the court's jurisdiction and "reimburse himself some $6,000 out of it."

Sherman's proposed settlement was drafted in the face of objections from Quick that $75,000 could not be obtained for the company's property. A delay would enable the leased Lower Division to yield $18,000 annually and the Upper Division $48,000 if leased also to the United Pipe Lines. The discrepancy between the lease rentals and the United Pipe Lines' proferred purchase price of $35,000 carried interesting implications. Sherman's plan was that consent decree judgments be entered awarding the trustees for the first- and second-mortgage bonds $72,000 and $53,740 respectively, together with a court order for the sale for a minimum of $78,000 of the franchises, pipelines, and other property which together secured the mortgages. When added to the $58,000 recovered from the Oil Creek railroad, total funds available could pay the judgments awarded the first- and second-mortgage holders, his own fee, and remunerate the receiver. The remaining balance could be distributed as a dividend to the general creditors of the company. Pending action on this proposed settlement Sherman took evidence

[31] Sherman Papers, Sherman to Brawley, April 14, 1879, Sherman to Wright, March 11, 1879, *Paper Books,* IX, No. 18, p. 36–38. *Appeal of Brown from F. W. Ames, Trustee.*

in preparation of the trial in the suit of Ames, Trustee, against the company.[32]

This negotiation demonstrated Sherman's skill in resolving complicated controversies within the field of equity. However, Wright and Kemble objected to the proposed settlement. Sherman then agreed to work with their Philadelphia counsel for a decree "for the amount of all the outstanding bonds," with the exception of bonds issued fraudulently. With reference to the latter Sherman's objection was in behalf of Ames, Trustee, for whom he was also counsel. As such he informed Wright's Philadelphia counsel that the lien of the first-mortgage bonds of the Pennsylvania Transportation Company did not extend to its Lower Division lines. Against those, the $1,000,000 bond issue represented by Derrickson, Trustee, had the prior claim.[33]

Sherman persuaded the master appointed to take testimony in the mortgage suits to decree foreclosure "for the full amount of the bonds" plus a lien upon the tolls and income (but not of the Lower Division lines) and to report at once so as to secure an immediate hearing of the cases in court. Although the master reported that Ames, Trustee for the first-mortgage bonds, held the "first lien on all property" the court stayed proceedings "for the present" on application of the receiver. Whereupon Wright asked Sherman if he couldn't secure a foreclosure "of some kind" in behalf of the first-mortgage bonds.[34]

Sherman must have found it difficult to serve the various parties that he represented in the complicated creditors' litigation with due regard to the interests of each group of clients and the requirements of the equity specialist's code. Very early in the protracted struggle to control the company's assets the general creditors had appointed him their counsel at a meeting when they instructed him to investigate the liability of Wright and Kemble and to proceed against

[32] *Ibid.,* Sherman to S. G. Thompson, June 18, 1879, Sherman to Kemble, April 25, 1879, *Paper Books,* VIII, No. 2, *Appeal of Brown,* Appendix, p. 45.

[33] *Ibid.,* Sherman to Thompson, June 30, August 1, 1879.

[34] *Ibid.,* Wright to Sherman, November 1879.

them personally for the entire amount of all claims against the company. During 1876 and 1877 Sherman had acted for these general creditors and secured judgments against the company for many of them for large amounts, while acting for Kemble and Wright and for Mackey in the attempt to secure a new receiver. In 1880, after Sherman had ceased to act as counsel of the general creditors, their new attorney, A. H. Bronson, pled with him wittily not to insist upon the full legal rights of Wright and Kemble, accused him of personal responsibility for the losses of the secondary group of claimants, and expressed bitter doubt as to how Sherman had fulfilled professionally the trust that they had originally bestowed upon him as their counsel. Sherman had been responsible as counsel for Ames, Trustee, for establishing the obvious priority of the first and second mortgages at the expense of the claims of the general creditors.

Most of Bronson's bitterness was provoked by Sherman's "smile of cool sarcasm" during a dispute in court.[35] The remainder derived from Sherman's success as Wright's counsel in defeating the general creditors' attempt to invalidate Wright's and Kemble's purchase for $79,999 of the Upper Division pipelines at the Crawford County Court of Common Pease decreed sale and also in recovering $60,000 from the receiver. Sherman represented those Philadelphia capitalists also in subsequent negotiations with S. C. T. Dodd and J. J. Vandergrift for the sale of the Upper Division pipelines for $60,000 to the United Pipe Lines, which company was already operating them under lease. That price was but a small fraction of the original capital investment that had made Harley and Abbott the leading pipeline operators in the Oil Region from 1868 to 1872. The sale, executed in July 1880, strengthened the Standard Oil's position in the Upper Division only moderately, since the United Pipe Lines and American Transfer lines already paralleled many of these pipes. Thereafter, only the Tide Water's gathering system offered competition to the Standard Oil's ubiquitous subsidiaries.[36]

[35] *Ibid.,* A. H. Bronson to Sherman, February 7, 1880.

[36] *Ibid.,* Sherman to "Dear Sir," February 23, 1880, Sweitzer to Sherman, February 27, 1880, Sherman to Vandergrift, March 20, 1880, Vandergrift

Shortly after this Sherman was retained by Quick to secure a rehearing of a Supreme Court order dismissing Quick as receiver in behalf of Harley and William Warmcastle, which action would have reinstated them in control of the now truncated company. This provided a gorgeous opportunity to castigate sarcastically the flagrant departures from business and legal ethics that had tricked the court into dismissing Quick, which Sherman capitalized upon with great effect and complete success. He disclosed how Harley's attorney, Joshua Douglass of Meadville, a director of the Pennsylvania Transportation Company, was acting also as counsel in the attempt of a James E. Brown to secure a judgment sale of the company's assets in his own behalf. This provoked a sarcastic reference to Sherman's other role as "counsel for Kemble and Wright" and the zeal with which he served them.[37]

The underlying issue in this contest was the Pennsylvania Transportation Company's suit against the Pittsburgh, Titusville and Buffalo Railroad for the large balance of the pipage owed the former by the Oil Creek railroad of which the latter was the successor. Sherman was invited into this suit as an additional counsel of the plaintiff when it was appealed finally to the Pennsylvania Supreme Court by Douglass and Brawley, who had been the opposing counsel in the preceding suit. While this attested to Sherman's reputation as the Oil Region's leading specialist in equity it illustrated also the lack of ill will that prevailed among members of the Crawford County bar. The trio of attorneys, in this case, exposed the collusion that had transmuted the Oil Creek railroad into

---

to Sherman, March 23, June 19, 1880, Thompson to Sherman, March 18, 1880, Sherman to Wright, May 8, 1880, Wright to Sherman, May 17, July 2, 16, 1880, Sherman to S. C. T. Dodd, June 19, 28, 1880, *Paper Books,* VIII, No. 2, pp. 70–72 and *passim,* "Decree of Court," *Appeal of Brown in Yarrow et al. v. Pennsylvania Transportation Company et al.; ibid.,* XII, No. 1, "Petition of Pennsylvania Pipe Line Company, October 5, 1881," *Pennsylvania Transportation Company v. Pittsburgh, Titusville and Buffalo Railway Company; Titusville Morning Herald,* March 17, 1880; Johnson, *op. cit.,* pp. 102–103.

[37] Sherman Papers, *Paper Books,* VIII, No. 1, *H. Harley and Wm. Warmcastle v. Pennsylvania Transportation Company.*

the Pittsburgh, Titusville and Buffalo Railroad and defended the sanctity of contracts with withering irony as they disclosed how the Pennsylvania Transportation Company and the Atlantic and Great Western Railroad had been excluded as creditors of the reorganized railroad. Sherman repudiated scathingly Judge Pearson Church's declaration in the Crawford County Common Pleas that in reorganizations, "by reason of the great magnitude of the interests of railway corporations they are not to be subjected to the ordinary rules of law." After a stinging denunciation of the chicanery that had fleeced railroad creditors in many recent operations of that type, he concluded with a demand that railroad reorganizations be required to observe both ethics and the law.[38]

While handling a variety of legal chores for Wright that arose from the sale of the Pennsylvania Transportation Company's Upper Division lines to the United Pipe Lines, collection of the purchase money from Vandergrift drew Sherman into a biting reply to S. C. T. Dodd, who had ridiculed his recital of the United Pipe Lines' departures from the sale contract's terms. Sherman answered bitterly, with oblique reference to the implications of the recently settled Commonwealth Suits: "I omitted references in the recitals to the 'fall of man' and the operations of the 'pardoning board' because I knew that your clients were familiar with the whole of the subject, and walking evidence of the fall of man."[39]

In the settlement of the Pennsylvania Transportation Company's affairs Sherman's hope for a 7% dividend for the general creditors, for which he and Heydrick planned some 50% would be contributed by the stockholders, went glimmering when the Brown estate under power of attorney collected 66.66% of the fund that he was gathering for this purpose and the Supreme Court of Pennsylvania upheld Judge Church's rejection of the equity bill against the Pitts-

[38] *Ibid., Paper Books,* XII, No. 1, pp. 3–8, 15, *Pennsylvania Transportation Company v. Pittsburgh, Titusville and Buffalo Railway Company,* Sherman's argument.

[39] Sherman Papers, Sherman to Wright, December 10, 1880, October 4, 1881, Wright to Sherman, December 1, 1880, Dodd to Sherman, December 9, 1880, Sherman to Dodd, December 10, 1880.

burgh, Titusville and Buffalo Railroad for the balance of pipage due. Heydrick bitterly called the exorbitant fees asked by lawyers representing the different claimants a "form of Kleptomania superinduced, as the doctors say, by insolvency of corporations and [one which] is alarmingly on the increase" [40] Sherman managed to reduce Quick's claim for extra compensation as the receiver, and received a fee of $1,000 from the master for services leading up to the sale of the Pennsylvania Transportation Company's remaining assets. From Wright and Kemble he received $4,000 late in 1880 for his services as their counsel in the final year of their litigation.[41] That in itself indicated his stature as a specialist in equity practice.

[40] *Ibid.,* C. Heydrick to Sherman, March 21, July 2, 5, 12, 1881, Sherman to Wright, September 24, October 14, 1881; *Pennsylvania State Reports,* CI, 576–583.

[41] Sherman Papers, Wright to Sherman, December 29, 1880, Sherman to Wright, January 6, 1881.

# CHAPTER X

# *A Regretted Change of Front*

BEFORE he received his fee from Charles B. Wright after the Pennsylvania Transportation Company litigation, Sherman was invited by H. L. Taylor & Company to join its counsel in a suit against the Standard Oil Company. The latter had succeeded in removing the case to the United States Circuit Court for trial when Sherman was called in to arrange a remanding of the case to the Crawford County Court of Common Pleas. The suit was one of a series arising from the quarrels of Taylor and J. Satterfield with John D. Rockefeller, Vandergrift, and others of that group over fulfillment of their partnership agreement. Since a part of that agreement provided for the marketing of the partnership's oil, the Standard Oil Company and its subsidiaries became involved in the widening controversy. Sherman consulted Shiras on the question of remanding, then decided independently that trial in the Circuit Court would have more advantages and persuaded Shiras to become consulting counsel. This case was settled out of court by negotiation. It illustrated further how Sherman's professional services were regarded as essential by business men, whether producers, oil buyers, or refiners, who became involved in equity litigation with the Standard Oil.[1] As long as such demands continued, they kept Sherman aligned with the business opposition to that business giant. Intellectually, of course, he was a Democratic antimonopolist.

This attitude had involved him during the Commonwealth Suits in 1879 in seeking independent antimonopolist journalism. First, in 1878 he urged his friend Whitman of the *Erie Observer* to come

[1] Sherman Papers, J. Satterfield to H. L. Taylor & Company, January 1, 1880, Sherman to Shiras, May 7, June 14, 1881; *Chicago Tribune,* November 28, 1879.

to Titusville and form a law partnership with him while launching "a Democratic *Weekly* for us." He went on to say: "The field is large. I have been so driven with business lately that I have scarcely time to eat." Whitman declined. Sherman's desire for an antimonopolist journal that would support the independent oil men persisted. This was enhanced by admiration of Henry Demarest Lloyd's widely read *Chicago Tribune* editorial, "The Vanderbilt-Gould Combination," of which he retained a copy in his papers. Sherman wanted a Sunday weekly that would avoid "the general lowness that distinguishes the Oil Region press." [2]

On May 31, 1880, after termination of the Commonwealth Suits, he joined forty-six Titusvillians and others in establishing a daily newspaper. They organized The World Publishing Company, Limited, with $25,000 capitalization. The resulting *Petroleum Daily World* was dedicated to keeping free enterprise and competition alive in Oildom. Sherman's subscription was a modest $500 among others ranging as high as $7,950. Among the subscribers were Campbell and Patterson, John L. McKinney, the local industrialist, the prosperous tailor John J. Carter, and ex-Mayor John Fertig. R. W. Criswell was editor, and James M. Place, the largest subscriber, business manager, an unwise choice. The *Petroleum World* began with great expectations as a vigorous anti-Standard Oil paper with twenty per cent of the subscriptions paid in.[3] Sherman was chairman of the Board of Managers. This was his second venture in Titusville journalism.

The *Petroleum World* fulfilled its anti-Standard Oil mission vigorously, even brilliantly, during its short life as a daily newspaper. Sherman contributed unsigned articles in an attempt to give it the tone that he desired. However, Place was an extravagant business manager who ignored the remonstrances of the Board of Managers. He paid nothing on his subscription, paid himself a salary dating from the canvass for subscriptions prior to organization,

[2] Sherman Papers, Sherman to Whitman, November 18, December 23, 1878, Sherman to Hoag, December 4, 1878.

[3] *Ibid.*, MS "Statement of The World Publishing Company Limited," May 31, 1880; Bates, *op. cit.*, pp. 318–319.

and purchased an unneeded large press. After Sherman investigated his management, the Board dismissed him in October. Sherman then did his best to persuade the disillusioned subscribers to pay up the balance of their subscriptions and to raise new capital, while reprimanding two former vice-presidents of the "Great Producers Council" for their attempts to repudiate their small subscriptions. The Board of Managers backed him by collecting defaulted subscriptions by legal action. He paid his own in full.[4]

With new capital, after absorbing the *Sunday Newsletter,* the *Petroleum World* reorganized. Able local correspondents in the Oil Region cities were hired. Sherman drafted the articles of association and the regulations, which Henry Byrom circulated among the subscribers. Sherman's attitude toward the venture he expressed to a reluctant subscriber: "I have felt many times that I was little better than a fool to spend my time and money in endeavoring to aid an enterprise for the public good, and without the hope or design of receiving any personal benefit, not even of getting back the money I had put into the company." Aided by his anonymous articles the daily *Petroleum World* continued until March 1, 1882. It crusaded for a free-pipeline bill, state and national railroad regulation, and antimonopolism in general, while championing the interests of the independent oil men. Then Frank W. Truesdell, foreman of the press room, formed a new company, purchased the paper, and continued it as a weekly, *The Sunday Petroleum World.*[5]

Henry Demarest Lloyd was the most prominent out-of-state subscriber. To him Truesdell confided important data on new developments in the oil industry in the Region. Although Sherman apparently severed connection with the *Sunday World* when Truesdell assumed control he contributed to it again in later years and in

[4] Sherman Papers, Sherman to Colonel J. A. Vera, September 6, 1880, Sherman to J. M. Fox, October 7, 1880, Sherman to Campbell, October 9, 1880, Sherman to E. E. Chambers and others, January 10, 1881, Sherman to F. H. Hall, January 18, 1881.

[5] *Ibid.,* Sherman to C. Fosky, February 4, 1881, Sherman to Beebe, February 4, 1881, Sherman to Henry Byrom and A. N. Perron, February 14, 1881, Sherman to Campbell, August 20, 26, 1881; Bates, *op. cit.,* pp. 318–319, 801.

this manner aided in reinvigorating the Jacksonian antimonopolism of the Oil Region, and at the same time resumed his correspondence with the more learned, cultivated Chicagoan.[6]

Elisha G. Patterson's subscription to the World Publishing Company in 1880 suggested the possibility of reuniting the producers' triumvirate that had played so extraordinary a role during 1878 and 1879. This was not to be. If that veteran antimonopolist could have revived his personal fortunes in a manner consistent with the independents' cause undoubtedly a reconciliation with Sherman and Campbell would have followed. In the spring of 1881 it appeared that Patterson might so recoup his fortunes while wreaking vengeance upon the Standard Oil Company of Ohio for its depredations upon the independents. The Auditor General of Pennsylvania, William P. Schell, invited him to assist as an expert in a pending tax suit for $796,642.20 against that foreign corporation in the Dauphin County Court of Common Pleas. As his reward Patterson would receive a fee proportionate to the sum in back taxes actually recovered up to a maximum of $23,000. The suit was based upon the principle that a foreign corporation was liable to the usual tax on stock proportionate to that portion employed in doing business in Pennsylvania. Patterson attempted to get Schell to go behind what the Standard Oil submitted to him as an agreed statement of facts and have the matter referred to a master in chancery for investigation.

Before Schell and Attorney General Palmer overruled this, however, Patterson sent Sherman as his counsel to New York in December 1881 to notify Standard Oil headquarters of this intention. There Sherman interviewed S. C. T. Dodd, now that company's chief law officer, and then John D. Archbold, who had risen to a high position in the ruling Standard Oil coterie. Sherman learned then that Patterson had sold out to Archbold secretly at some earlier date, probably in connection with the initiation of Commonwealth v. Rockefeller *et al.* Archbold impugned Patterson's motives in

[6] Lloyd Papers (State Historical Society of Wisconsin), Sherman to Lloyd correspondence.

threatening to go behind the agreed statement of facts that he had supplied to Palmer and Schell in the tax suit, and intimated that the Standard Oil would employ that antimonopolist if he withdrew from the tax suit as expert.

Sherman immediately refused to negotiate as Patterson's counsel and withdrew from all connection with the tax suit. After his return home, and after receiving suspicious telegrams from Patterson, Sherman cut him on the street. He never recognized him thereafter although he saw him frequently in the city for some years. This was an effective method of informing the independent oil men of what had happened without violating professional obligations fo a former client.[7] Nevertheless, discouraged by Patterson's earlier secret betrayal, Sherman became susceptible to an invitation that he join the Standard Oil's legal staff.

Several months after this episode in Titusville, Patterson's new attorney, S. T. Neil, left a conference between Patterson and Schell and Palmer at Harrisburg and went to Titusville. There he called upon Sherman, whom he and Patterson knew had contracted to join the Standard Oil's legal staff. Neil informed him that Patterson was about to have proceedings instituted for taxes allegedly due to the state from the United Pipe Lines and American Transfer companies, which Schell had compromised away. Neil intimated that this would be done on the following Tuesday and "that it was advisable to secure to P. whatever commissions might be in this for him." Sherman replied that he would have nothing to do with this, a hint that he knew that Patterson was maneuvering for another bribe from Standard Oil sources. He notified Archbold immediately of this approach and learned that that magnate was agreeable.[8]

Eventually the trial court ruled that only $33,270.59 in taxes and penalties were due from the Standard Oil, which the Pennsylvania Supreme Court cut to $22,660.10 on appeal. Indignant, his compen-

[7] Sherman Papers, E. G. Patterson to Sherman, January 16, 31, 1882; *Report of the Proceedings before the Committee Appointed by the Legislature of Pennsylvania to Inquire into the Legal Relation of the Standard Oil Company to the State,* pp. 535–536, testimony of Roger Sherman.

[8] Sherman Papers, Sherman to John D. Archbold, February 27, 1882.

sation as expert reduced to a pittance, Patterson attempted to reopen the suit before the Supreme Court by offering new evidence to be presented to a master, whose appointment his counsel petitioned for after presenting his affidavit as to the facts. Palmer declined to join in the action to reopen the tax suit, but invited Patterson instead of consult him at Philadelphia. There, in the same hotel, Archbold appeared and called upon Patterson, only to learn that he would still not be bound by the agreed statement of facts as to the Standard Oil's assets in Pennsylvania, upon which the proceeding had been based, and that he intended to proceed with his action. Archbold then proposed that Patterson accept a sum of money ($15,000) equal to the commission that he would receive from the Auditor General if he succeeded in enlarging the statement of Standard Oil assets in the state, and also a $5,000 a year position with the Standard Oil. Patterson accepted the arrangement. With some of the money paid him by the Standard Oil for desisting from his attempt to collect a larger sum of back taxes for the state he paid up his subscription to fifty shares of stock of the Tide Water Pipe Company, to which he had subscribed ostensibly as a leader of the independents.

Patterson immediately instituted charges against the Tide Water management, alleging mismanagement and insolvency in a petition to the Crawford County Court of Common Pleas. Benson's and McKelvy's counsel put him on the witness stand on December 20, 1882, and extracted from him an admission of what had happened at that secret Philadelphia conference. This caused a scandal. Many inferred that Archbold had bribed him to enable the hated Standard Oil to evade taxes justly due. The scandal was greater because Patterson admitted that he had "had business relations" with that company "during the last three years," beginning apparently during his participation in the prosecution of the Commonwealth Suits. After the state election of 1882 a joint legislative committee investigated the matter fully. Before it, Franklin B. Gowen established the *prima facie* presumption, which Archbold denied, that the Standard Oil Trustee had bribed Patterson to desist from what would have been a successful attempt to reopen the tax suit and collect more back taxes from the Standard Oil despite the opposition of Attorney General Palmer.

Gowen's dramatic questioning of Archbold and Patterson, and the accusation of Henry M. Flagler, Secretary of the Standard Oil Trust, that Patterson was a blackmailer, destroyed the latter's reputation among the independent oil men. They regarded him thereafter accurately as a Standard Oil tool. He continued in that organization's employ, and moved eventually to Independence, Kansas.[9] Sherman's and Archbold's private knowledge that Patterson had invited the bribe in the tax suit was not divulged in the investigation. Flagler's accusation was entirely true. Had the independent oil men's movement not virtually disintegrated before this partial revelation, Patterson's betrayal would have dealt it a mortal blow.[10]

Patterson's attacks upon the settlement of the Commonwealth Suits of 1880 at the General Council of the Producers Unions had provoked wide criticism of Sherman and Campbell, who had negotiated it. Some producers concluded that they had yielded to corrupt influences. Those attacks appeared now in retrospect to have originated with Patterson's "business relations" from 1879 to 1880 with the Standard Oil. Sherman's attempt to perpetuate the *Petroleum World* as an anti-Standard Oil organ from 1880 to 1882 vindicated his loyalty to the independents in the complicated settlement of 1880.

Why Sherman decided early in 1882 to accept Archibold's invitation to join the Standard Oil's legal staff on a five-year contract does not appear in his surviving correspondence. Probably Archbold made this offer in Titusville at Christmas time 1881, after Sherman's open break with Patterson, when the Titusville lawyer was stunned by the discovery of his former client's earlier betrayal of the independents and discouraged by the decline of the independent

[9] *Pennsylvania State Reports,* CI, 119–151, Commonwealth *v.* Standard Oil Company; Lloyd, *op. cit.,* pp. 166–180; Sherman Papers, MS (copy) E. G. Patterson's testimony, December 20, 1882, in E. G. Patterson *v.* Tide Water Pipe Company; *Report of the Proceedings before the Committee Appointed by the Legislature of Pennsylvania, passim,* but especially Gowen's opening argument and the testimony of Archbold and Patterson.

[10] *Titusville Morning Herald,* April 13, May 12, 1880.

movement and the bad fortune of the *Petroleum World.* Before accepting, Sherman sought vainly a conference with Whitman. He did not discuss Archbold's invitation with Beebe, his former law partner, or with E. M. Guthrie, a powerful antagonist during the Commonwealth Suits. Apparently Sherman weighed Archbold's offer for some weeks before accepting it in February 1882. After he did so, Guthrie informed Beebe that the salary stipulated in the contract was $20,000. In his memoir Sherman stated that it had been $12,500 annually.[11]

Early in 1882 his professional income from independent oil sources, which patronage he preferred and with which he was intellectually and psychologically compatible, was declining. His law practice was suffering otherwise from the continued decline of oil production in the Oil Creek district. While the independents were diminishing in wealth if not in number the Standard Oil's continued dominance of the industry revealed the limitations of the former's victory during the settlement of 1880. With his faith in the ability of the independents to survive drastically diminished, disillusioned by Patterson's betrayal, lacking in substantial means, and obliged to support his family from his law practice, Sherman accepted what was obviously to him an unpalatable professional engagement, probably in the hope that it could be made agreeable.

Archbold's motives must also be analyzed. Rockefeller with Dodd's aid was then reorganizing his petroleum business empire into the Standard Oil Trust, which was secretly launched on January 2, 1882. It was Rockefeller's undoubted objective to remove from the legal lists so redoutable an antagonist as Roger Sherman. Since Dodd had crossed swords with him in the courts repeatedly he knew the caliber of his steel. Rockefeller usually preferred to take a formidable competitor or antagonist into his organization rather than to fight it out with him to the bitter end. Although Sherman lacked large resources to finance an effective counterattack, if he discovered the existence of the new "Trust," Rockefeller

[11] Sherman Papers, Sherman to Whitman, February 28, 1882, Beebe to Sherman, March 2, 1882, Sherman to G. A. Vilas, August 31, 1885; Sherman, *op. cit.*, p. 73.

regarded it as wise to add his recognized talent to Dodd's legal staff before he did so. This would and did prevent an attack upon it for five years by the independents' ablest counsel.

Soon after Sherman signed the five-year contract the press learned that he had become a Standard Oil attorney. This discovery precipitated a bitter press attack upon him that was capped by an article in *The Oil Paint and Drug Reporter* containing especially derogatory allusions and charges impugning Sherman's professional integrity. These attacks were widely believed and were regarded as confirming Patterson's insinuations in the debate in the General Council of the Producers Unions after the negotiated settlement of 1880. So bitter were these attacks that Joshua Douglass, a Meadville attorney, informed Sherman of a rumor that he was about to leave for "greener pastures," a hint that he would be wise to do so. Sherman sent his friend Whitman a collection of the newspaper articles. The article in *The Oil Paint and Drug Reporter,* he declared, "does me a great injustice. I trust I am not such a fool as to do anything at this late day that the most critical could construe into a violation of professional duty, and I certainly have not done so in this instance."[12]

Sherman had abandoned private practice for the position of a salaried attorney on the staff of S. C. T. Dodd, General Solicitor of the still secret Standard Oil Trust, with whom not long since he had exchanged quips about "the fall of man." If Dodd's superiors were examples of that phenomenon as Sherman had charged, certainly he had joined them when he abdicated as the intellectual and legal leader of the independents' long opposition to the Rockefeller organization.

Many of Sherman's former clients and admirers among the independents regarded him as a fallen hero. For years afterward it was believed and to the present day there persists among some people in Crawford County a conviction that Sherman had betrayed them. Lewis Emery, Jr., did not believe this, and retained Sherman

[12] Sherman Papers, Douglass to Sherman, February 28, 1882, Sherman to Whitman, February 28, 1882.

again later as his counsel when he resumed private practice. Now out of debt and increasingly successful in the Bradford field, Emery recognized that an attorney enjoyed the legal right to change his clients. However, in view of what Sherman had disclosed of Standard Oil business and political methods in the Commonwealth and Ohlen suits, it may be asked if he did not violate the obligation of the specialist in equity to refuse to represent or advise as counsel corporations and businessmen of whose business methods he disapproved.[13] He could reply that the Standard Oil had contracted to reform its methods.

Like many another believer in competitive free enterprise and effective business ethics Sherman had accepted the Rockefeller yoke, knowing that magnate had been the bitterest enemy of both. Sherman ignored publicly the widespread criticism of his changed professional affiliation. He held this to be a "matter of sentiment entirely personal" to himself. He took into his office a Samuel Minor to whom he delegated most of the routine Standard Oil legal business assigned to himself.[14]

Sherman handled personally the more important legal chores assigned him by Dodd. For example, he negotiated quietly for the Standard Oil the purchase of the Solar Oil Works of Buffalo, a defeated refining competitor which had introduced Samuel Van Syckel's pioneer invention of continuous distillation. In July 1882 Archbold informed Sherman and Bloss, (editor of the *Titusville Morning Herald*), who the *Oil City Blizzard* declared had taken Archbold into partnership, that the Standard Oil would approve if the Titusville attorney ran for the legislature. If Sherman did so, Archbold also informed him, Dodd would adjust his legal work accordingly and the Standard Oil would lend its fullest influence to secure his election. As Sherman viewed it, this was a polite way of informing him that there was another way in which he could serve the Rockefeller organization while remaining on the payroll. It also confirmed his earlier inference as to the basis of the Standard Oil's political influence at Harrisburg. This advice Sherman received

[13] Cf. Walter Jessup, *Elihu Root, I* (New York, 1938), 80–81.
[14] Sherman, *op. cit.*, p. 73.

before the trial of Patterson *v.* Tide Water Pipe Company, which he did not attend.[15]

After a year's service under Dodd, during which time he became increasingly dissatisfied, Sherman offered his resignation. To his great surprise, Rockefeller and the other Standard Oil Trustees held him to fulfillment of his contract. However, they allowed him to resume general practice also, while retaining him at the same salary in their interest. It was soon evident that despite this generous arrangement Sherman resented being held to the contract and the unpalatable service required by it. Neither Rockefeller nor Dodd wished to release him at a time when the producers were experimenting again with the shut-down of wells as a method of controlling supply and evincing revived faith in their ability to defeat the Standard Oil's control of the market by concerted action. After Rockefeller rejected his resignation and altered his connection with the Standard Oil to that of attorney on retainer, Sherman informed George Shiras, Jr., one of his few remaining close professional friends of this. Shiras replied: "I am glad to be informed that you have reached an understanding with the Standard people consistent with pecuniary interest and professional honor. . . . I shall take pleasure in drinking your health in some of the champagne which was duly left at my house." [16]

Sherman could hardly have enjoyed testifying under Dodd's aegis, December 18, 1883, in Philadelphia during the congressional Committee on Commerce's investigation of the Standard Oil. Neither did Sherman enjoy having to act as counsel of the minority Standard Oil stockholders of the Tide Water during their struggle with the latter's management. Benson and he had been close friends.[17]

[15] Sherman Papers, Archbold to Sherman, June 7, July 10, 1882, January 13, 1883, Archbold to Bloss, July 10, 1882, Sherman to Archbold, August 1, 1882, W. S. Rose to Sherman, December 20, 1882.

[16] *Ibid.,* March 15, 1883.

[17] *Ibid.,* Dodd to Sherman, October 27, December 10, 1883, September 13, November 14, 1884, Sherman to Dodd, January 16, September 6, 13, October 7, November 12, 1884; *Pennsylvania State Reports,* CVIII, 630–637.

Sherman maintained cordial professional relations with Dodd and Archbold. When the opponents of speculation in oil certificates asked him to draft a bill for presentation at Harrisburg and Albany, to "restrain and greatly prohibit" such speculation, he consulted Dodd in friendly fashion. During the following February he attempted to enlist Archbold's aid in securing appointment of a relative of a pro-Standard Oil stockholder of the Tide Water as United States District Attorney for the Western District of Pennsylvania. On another occasion he invited Archbold to support the candidacy of N. M. Allen "for the mission to the Argentine Republic." To Captain Vandergrift, President of the United Pipe Lines, he loaned a volume containing the charters of the "old Penn. Transp. Co.," enjoining him to take good care of the book and return it when he had read it.[18]

The available record of Sherman's five years' service with Dodd's legal staff for the Standard Oil is slender. Most of the legal papers accumulated during that period he returned to Dodd on expiration of the contract. Since every request by the present author for permission to examine the archives of the Standard Oil Company of New Jersey has been refused, the account of this portion of Sherman's career cannot be expanded. The available evidence indicates that he did very little work for the Standard Oil Trust and its subsidiaries.

In March 1887, when the Standard Oil's attorney in Buffalo sought his assistance in meeting Samuel Van Syckel's suit for damages arising from infringement of his invention of continuous distillation Sherman replied that his connection with the Rockefeller organization had ceased on February 28. Dodd was notified immediately. He wrote promptly to Sherman: "Our attention had not before been called to the fact that your agreement had terminated. We desire to retain you in our business, and I have been requested to correspond with you with the view to making some satisfactory business arrangement." He suggested "a satisfactory annual retainer, you to charge your usual fees for such business as you may do,"

[18] Sherman Papers, Sherman to Archbold, February 19, June 5, 1885, Sherman to Vandergrift, February 28, 1885, Sherman to Dodd, November 10, 1884.

and continued: "Please let me hear from you at once, as we must engage some one to take testimony in the Van Syckel case this week unless you act for us."[19] This was not in accord with Sherman's preference. He refused to continue in the Standard Oil's service and resumed his independent practice.

Possibly the desire to have him defend the Standard Oil in Van Syckel's suit may have precipitated this decision. Sherman had represented him years before at the beginning of his practice in the Titusville area, and he may have anticipated the favorable verdict that the inventor won in the courts in this new action. However, Sherman had long strained at the Rockefeller leash, and returned to the uncertainties of private practice with rejoicing. He knew that he might be handicapped by dislike of his recent Standard Oil connection in the anti-Standard Oil section of the Oil Region's business community. That Sherman had abandoned the lucrative position of being a Standard Oil attorney on retainer with all the prestige that that brought in railroad, "Big Business," and high political circles spoke volumes to his contemporaries.

Despite his service with the Standard Oil, Sherman had asked Shiras early in 1885 to help him to secure the appointment of commissioner in bankruptcy for Crawford, Butler, and Mercer counties under a new law that the legislature was expected to enact. In May 1887 after he had become again his own man, Sherman went to Washington. There he called upon Congressman Samuel J. Randall, the influential Pennsylvania Democratic leader.[20]

While this confirmed his regularity as a Democrat, the *American Citizen* which he had launched in January 1885 had enhanced the Democratic party's strength in northwest Pennsylvania. It also publicized Sherman's views on important issues.

The *American Citizen*'s first issue on January 3, 1885, had dedicated that weekly to the larger interests of the area. Recognizing that "the petroleum business in many of its branches has passed away and is passing away from our immediate locality," Sherman

[19] *Ibid.,* Dodd to Sherman, March 21, 1887.

[20] *Ibid.,* Sherman to Shiras, February 19, 1885, Sherman to J. B. Brawley, May 19, 1885.

fostered local industry, trade, and agriculture. His local objective was to foster diversification and recovery of Titusville's economic life, whose decay had been depicted vividly two years before by the new Merchants and Manufacturers Association.[21] Apparently, Sherman's light schedule as a Standard Oil lawyer had left him with ample leisure which he could devote to editing the new weekly.

The leading editorial of that first number, "The Democratic Faith," hailed the Democratic party's national victory of 1884. This, Sherman predicted, would inaugurate an era of reform. It would restore honesty, simplicity, and regard for private rights in place of the former Republican alliance with money and aristocracy and its rising tide of misconduct and corruption. Civil Service Reform would end bribery of the electorate and official corruption. Public lands would be redistributed to home owners after forfeiture of unclaimed railroad land grants. Business would be restored to "normal conditions" untroubled by "government tinkering and favoritism." The Negro had nothing to fear from Cleveland and would be "brought to know the obligations of his higher position." Thus, Sherman announced his identification with moderate reform's opposition to political Bourbonism and privilege-seeking business interests.

There was no evidence that the last eighteen months of his service with the Standard Oil influenced Sherman's editorship. That was distinguished by intellectual and political opposition to the system of business and politics which the Rockefellers did so much to strengthen. From that system Sherman would free northwest Pennsylvania as he attacked it in his next issue for perpetuating Civil War issues and fostering race hatred, sectional division, and conflict. In the North "the Republican Bourbon is the dangerous element," he declared. Reunion and mutual respect between North and South was the proper policy. Turning from the past Sherman stressed the importance of preserving constitutional democracy, binding the Union together, ensuring liberty and order, and protecting the ballot from bribery.[22]

[21] Sherman Papers.

[22] *American Citizen,* January 10, 1885.

H. C. Eddy & Company (Henry C. Eddy and Roger Sherman) published the *American Citizen* and operated a job printing plant. Advertisements in the weekly were almost entirely those of local industries, merchants, and professions. By December 1885 the paper had expanded from four to eight pages. It served the purposes that Sherman had asked his friend, Whitman, to cultivate six years before. While fighting for the Cleveland school of reform the paper combined the function of a literary vehicle with editorial campaigning, local, regional, and agricultural news, and brief items of general interest from domestic and foreign exchanges. The advertising and job printing enabled the firm to balance its books, if not to show an actual profit. The praise that the new journal received indicated Sherman's competence as a journalist, while the paper itself provided insight into his intellectual, political, and reform interests.[23] He exerted editorial influence in behalf of desirable Democratic candidates in the Oil Region, where his party continued to exhibit substantial strength.

As a literary periodical the weekly was decidedly second rate, not comparable with John McGovern's (Chicago) *Current,* which combined critical comment upon contemporary problems with acute literary criticism and much better literary fare. The *American Citizen*'s first page was filled, at first, with fiction and verse. The original contributions were frequently anonymous and devoid of great merit. They were relieved irregularly by republished short works of leading writers, such as Charles Dickens' "The Redeemed Profligate." [24]

The editorials, by contrast, were of high quality in form and content. Many provided readers with a well-defined political philosophy, showing the influence of the National Civil Service Reform League and of Henry Demarest Lloyd. The former's cause was presented regularly in a manner resembling that of Theodore Roosevelt on the Civil Service Commission. Lloyd's influence reinforced

[23] Sherman Papers, Eddy & Company to Publishers of *Century* and *North American Review,* December 26, 1885, Eddy & Company to E. M. Guthrie, April 26, 1885.

[24] February 7, 1885.

Sherman's experience and the Jacksonian tradition in shaping his attitude toward corporations. In an early editorial he developed the maxim: "Corporations have no souls, and their managers often have no sense." Sherman declared the recent consolidation of the New York Central Railroad with the West Shore was "illegal" while reiterating Jeremiah Black's thesis that the public interest should be paramount in railroad management. The clause in the Pennsylvania Constitution prohibiting purchase or lease by one railroad of another he observed was "a direct blow at monopoly." The public was entitled to the benefits of competition, he added, voicing another shibboleth of the economic faith of Oildom.[25]

Opposition to corrupt bosses and political rings Sherman combined with championship of the Democratic party as a reforming, liberating agency. He attacked Matthew S. Quay repeatedly in August 1885 when he was candidate for State Treasurer, as having been "a political boss, ringster and trickster for the greater part of his life." His record proved that he was a "favorite of the Pennsylvania Railroad," the confidant of Kemble and Salter in the attempted $4,000,000 "Riot Bill steal" of 1878, and responsible as a member of the Pardon Board for pardoning Kemble after he had been sentenced to the penitentiary for flagrant corruption of the General Assembly. The real issue of the state election of 1885, Sherman declared, was between "the will of the people" and the Pennsylvania Railroad and Quay, its "representative of jobbery." The public will, as defined by Sherman, combined antimonopolism with equalitarianism in a significant revival of the Jacksonian faith.[26]

As he approached the end of his contract service with the Standard Oil, Sherman editorialized cogently in January 1887 on "The Growth of Classes." Among the "distinct classes" he identified the capitalists as being "by far the most pushing and influential with their lobbies and demand for special legislation at odds with the general welfare." Other classes were the national bankers, patent holders, war veterans, and protected manufacturers, illustrating a tendency toward social stratification that was demonstrated also by

[25] August 21, 1885. [26] August 28, October 16, 1885.

admiration of titles and attempts to perpetuate class distinctions. Thus Sherman published his awareness of a significant trend of the era. His reaction to it was characteristically Jacksonian. No law, he declared, should benefit one class more than another. He asserted the principle of equal, uniform laws as the proper legislative maxim.[27]

During that same January Sherman proclaimed his adhesion to the moderate regulatory theory when he hailed the Interstate Commerce Act. This, he declared, was "one of the most important laws enacted in the history of the United States," the need for which had been apparent for fifteen years. It would, he declared, be practicable "to do business without the pooling and rebate system." The Interstate Commerce Act, he added later, "is the greatest step in this country in the present century towards emancipating the people from the unjust exactions of railroad corporations and monopolies." However, he said: "These have all combined in a great conspiracy to render the law incapable of enforcement." With this "in mind" the public should "stand by the law." [28]

On March 4, in the first issue after his return to full private practice, he declared that the United States Senate was "made up of very rich men, only a few of them having abilities above the average while the wealth of many had been acquired by methods that would not bear investigation." In the national House of Representatives he saw "a constant inflowing mob of new men distinguished for ignorance of public affairs and lack of experience." Obviously, Cleveland's administration had neither terminated Bourbonism nor ushered in the "Golden Age" in public life. In subsequent issues Sherman declared that the public was weary of "Bloody Shirt" partisanship and that James G. Blaine had "not developed a moral principle of action" during twenty-five years of public life. He criticized the demands of war veterans, denounced the political power of the saloons, analyzed illegal capitalistic combinations and unpunishable conspiracies of capital against labor, exposed the venality of the press, and depicted the prevailing spoils system and

[27] January 7, 1885. [28] January 28, April 22, 1887.

corruption of voters. He decried the political apathy of the middle class and the situation that obliged it to choose between tickets named by party bosses.

Among many issues that the *American Citizen* stressed, the most significant was the rising antipathy to monopolies. It would require long, patient work to undo the effects of "indirect assaults upon the principle of equal rights and open trade," Sherman declared. He opposed speculation on all exchanges in a manner similar to Lloyd's *Chicago Tribune* editorials. Sherman then referred to "Imaginary Petroleum" in an attack upon speculation in oil certificates.[29]

During that same April Sherman remarked realistically that the Associated Press was "controlled in part by subsidized organs of the railway and other corporations." It was "allied to and does the bidding of Jay Gould's Western Union Telegraph Company" and was aiding Gould in his attempt to discredit the Interstate Commerce Act while monopolizing, distorting, and censoring the actual news. This was an issue that Henry George had raised in California a decade earlier. That he was aware of the social injustice produced by the Second Industrial Revolution and its sweatshop industry he had evinced in February by republishing Hood's "Song of the Shirt" from *Puck*.[30]

Sherman enjoyed his editorship. By this time the *American Citizen* exhibited his skill in catering to a regional clientele. News articles and paragraphs shared the front page with a series of articles on the famous hostelry, "The Mansion House," now the Colonel Edwin L. Drake Hotel of Titusville. Literature composed of original and reprinted fiction, original and reprinted verse, humor, and continued stories filled the next two pages. Then "Neighboring Correspondence" from surrounding towns and villages filled page four together with "Titusville Market" reports on produce prices for the benefit of farmers. Page five was devoted to editorials and Titusville news. "For the Farmers" filled page six. Crawford County official news and the featuring of unusual persons and stories fol-

[29] March 11, April 15, May 6, 1887. Cf. Sherman Papers, "Index to 'American Citizen,' " I–V.

[30] February 25, April 22, 1887.

lowed. Page eight contained a condensed "Epitome of the Week" of congressional, American, and foreign news. Since no price per copy was quoted it was apparent that readers had to subscribe to the weekly on an annual basis.[31]

Work on the *American Citizen* filled hours that would have been idle otherwise until Sherman's law practice revived. He suffered severely at first from reduced income after he declined to continue as a Standard Oil attorney. This sacrifice was authentic evidence of his attachment to antimonopolist and competitive principles. Evidently he hoped to attract again to his practice clients from the petroleum industry with similar views, who were interested in competing with the Standard Oil. He continued as editor of the *American Citizen* until late in 1889 when his practice once more demanded his entire working day. He then sold his interest in the paper to Eddy, who continued its publication on his individual account.[32]

Incidentally, while Sherman provided the Democratic party with a local organ in Crawford County during those five years, he did not succeed in rallying sufficient strength to it to overturn the Republican majority. The *Morning Herald* served this group efficiently but with increasing conservativism that bespoke the secret Archbold partnership in its management. In November 1888 Benjamin Harrison carried the county by 8,040 votes to 5,964 for Cleveland.[33] The protectionist Bourbonism of Quay's political machine was as yet too powerful for the moderate political reform that Sherman advocated.

[31] February 25, 1887.
[32] Bates, *op. cit.*, p. 319.
[33] *American Citizen*, November 23, 1888.

# CHAPTER XI

# *Joining the Civic Elite*

CIVIC and cultural leadership in Titusville continued to be influenced largely by the Presbyterian Church. The upper strata of the city was largely Presbyterian. Although Roger Sherman had been baptised in the Episcopalian Church he did not join St. James. Instead, after Mrs. Sherman joined the First Presbyterian Church on April 4, 1873, he attended services with her there. They enjoyed the friendship of the influential Presbyterian business and professional families. Probably, since he did not join that church, Sherman's war experience and subsequent reading had made him sympathetic with Robert G. Ingersoll's religious liberalism, judging from his collection of that attorney's publications. When Robert E. Hopkins of the Tide Water Pipe Company became chairman of the music committee of the First Presbyterian Church, Sherman twitted him gaily. He wrote Hopkins that he was the first chairman to complain against the custom of paying the salaries of the choir from his own pocket. "I think the Committee. . . should hold you strictly to the performance of all the duties." Sherman added: "Still, as you seem to have been under a misapprehension. . . I enclose my check for $12.50 for the first quarter of my subscription." [1]

Although he had his own private library Sherman was interested in establishing a library in the city. When the organization of libraries in Meadville, Erie, and Warren made it possible to appeal to local civic pride, he enlisted the support of the Presbyterian pastor and others. Then he convoked a meeting of interested civic leaders in his office in March 1877 and informed the group that it

[1] MS Minutes of Session and MS Record of Members, First Presbyterian Church of Titusville (courtesy of the Rev. Richard S. Graham); Sherman Papers, Sherman to Hopkins, April 21, 1879.

was high time the city established a permanent library. A bipartisan Citizens Committee to push the project was headed by Sherman and included among its members N. M. Allen, David Emery, and Bloss of the *Titusville Morning Herald.* By April 1 the Committee had raised $2,400. At a meeting in the Sherman & Beebe law office Sherman proposed a chartered subscription library with a public reading room. A. E. Perrin urged that it be open "to all classes" and "catch the men on our street corners." Sherman headed the guarantors' subscriptions to single shares of stock of $100.

At a stockholders meeting April 3 Sherman reported the plan of organization. This charter specified that all income above maintenance and expenses should be expended on books and library improvement. The stockholders were made liable to pay periodic assessments for both purposes. The sale of membership tickets at $2 a year and the income of special benefits were to provide most of the funds to maintain the reading room, pay the librarian, and purchase books and periodicals. The constitution and bylaws forbade indebtedness and required closing the library if insufficient membership cards were sold. David B. Benson was elected President and Sherman Secretary of the new Titusville Library Association. As Secretary Sherman was an important factor in the library's success. For some years he devoted his leisure time to ordering the books and periodicals that provided the reading room with a growing collection, to conducting the Directors' correspondence and to recording their proceedings.[2]

Eight months after opening he reported on May 1, 1878, a circulation of 500 books per month and full use of the reading room by high school youth, "mechanics and clerks" and others. Since only 200 members subscribed that first year Sherman persuaded heads of families and others with means to give memberships to young people. The proceeds of public lectures, lawn parties, and other benefits contributed additional funds. In June the Board acquired 1,800 volumes of the defunct Temperance Union

[2] Titusville Public Library, MS "Record, Titusville Library Association," MS Letter Press Copy Book, MS "Cash, Titusville Library Association"; *Titusville Morning Herald,* March 12, 13, April 2, 4, 1877.

Library and increased periodicals to 146. In October the *Morning Herald* complimented the patronage of "the noble library," whose crowded reading room and large circulation it declared was an honor to the founders. When the Shermans opened their residence in September 1879 to a benefit "musicale" the *Morning Herald* urged friends of the library who were not of the upper social elite to attend. It encouraged the committee of ladies which was selling membership tickets. Money for new books was also provided by annual assessments on the stockholders who paid them "freely and without hope of reward." [3]

Inadequate sale of membership tickets necessitated dependence upon special benefits, which contributed to the city's cultural life. The program of 1880 included an opera-house series of readings by Mrs. Adele Bross, a play by the Titusville Dramatic Association, a Leap Year Party, a Mendelssohn concert, a Philomel Lady Quartette concert, and a lecture course that netted some $1,000. By that time the book collection had increased to 3,000, its excellent character attested by a printed *Catalogue.* Both independent oil men and local Standard Oil officials were represented among the stockholders. Among the former were Sherman, David McKelvy, Benson, David Emery, and Henry Byrom, while John D. Archbold and Quick represented the latter. Willian T. Scheide, district manager of the United Pipe Lines Company, who was collecting a unique private library, did not join them.[4]

With the decline of the city's population in the early eighties the annual sale of membership tickets declined also. The loss of fifteen stockholders by death or removal diminished the guarantors' ability

[3] Sherman Papers, "Titusville City Library. Its Condition and Needs," unidentified newspaper clipping, Sherman to E. A. Potter, October 25, 1879, Sherman to Mrs. L. A. Fowler, February 5, 1879, Wm. J. Carpenter to Sherman, March 12, 1879; Titusville Public Library, Titusville Library Association, MS "Ledger"; *Titusville Morning Herald,* September 26, 1879.

[4] *Catalogue of the Titusville Library Association* (Titusville, 1880), pp. 3–4, 7–141; Titusville Public Library, Titusville Library Association, MS "Ledger"; Sherman papers, Sherman to R. M. Streeter, December 31, 1879, Sherman to Hopkins, August 16, 1880.

to sustain the library. Sherman was elected President unanimously in April 1885. He secured rent-free quarters where a part-time librarian administered the library for $200 annually. Sherman was re-elected in 1886 when the book collection had risen to 7,412 and memberships to 253. In March 1887 the Directors contracted with the Young Men's Christian Association to house and administer the library for three years. Sherman again became a Director and was elected Treasurer in 1890, positions that he held for some years. The library was removed to the High School where it remained until it was placed in the front rooms of the City Hall in 1896. The failure to finance the library adequately on a subscription basis paralleled the experience of neighboring cities. The Woman's Club of Titusville, of which Mrs. Sherman was a President, was especially interested in its perpetuation.[5]

In August 1883 Roger Sherman was elected, on the nomination of R. Francis Wood of Philadelphia, a member of the Civil Service Reform Association. He preserved the publications of that crusading organization of the civic elite. This inspired him to attempt to terminate the reign of party bosses and rings in Pennsylvania's political life. Among these pamphlets was *The Spoilsman and Civil Service Reform,* which quoted Charles J. Bonaparte's denunciation of the immorality of the spoils system whose threat to "free institutions" he declared that civil service reform would terminate.[6]

In 1881 as a member of the Resolutions Committee of the Democratic State Convention, Sherman drafted the platform resolutions that opposed the aggressions and oppressions of corporations, whereby they sought to establish monopolies. When the state Republican majority was reduced the following November to 7,000 the Republicans attributed this to the "radical character" of the Democratic platform.[7] In 1882 Sherman served a year as President of the Young Men's Democratic Club of Titusville, and promoted

[5] Titusville Public Library, Titusville Library Association, MS "Minute Book No. 2," pp. 3–6, 9, 16, 28–29, 45–51, 56, 63–69, 86–89; *American Citizen,* December 9, 1887; *Titusville Morning Herald,* March 31, 1892.

[6] Sherman Papers, *Pamphlets,* XLIV, No. 7.

[7] Sherman, *op. cit.,* pp. 72–73; Bates, *op. cit.,* p. 801.

the re-election of Justice Trunkey to the Pennsylvania Supreme Court by attracting to him the anti-boss support of independent Republicans.[8] In 1884 Sherman was almost elected Mayor of Titusville, which was normally Republican. In 1888 he probably helped to prepare the antimonopolist pamphlets that the Democratic State Committee circulated among farmers and coal miners attacking pools, trusts, and land monopoly. These he preserved carefully. He was nominated for delegate-at-large by the Democratic State Convention to the proposed constitutional state convention, which was defeated in the autumn election.[9]

In 1890 Sherman hired private detectives to expose and prevent continued Republican bribing of voters. This contributed significantly to substituting a 6,000 Democratic plurality for Robert E. Pattison, which elected him Governor, for Harrison's 10,000 plurality in 1888. Sherman then put himself at the head of the "northwest movement" and demanded that the Governor-elect select a cabinet member from northwest Pennsylvania. He was himself strongly supported for Attorney General but to preserve unity urged all claimants to join him in supporting his friend Heydrick for that post. The movement disintegrated when Pattison refused to make the cabinet appointment.[10]

Usually Sherman sought no political office. In 1887 he desired appointment to the Interstate Commerce Commission to help subject the railroads to legal restraints in behalf of fair business competition. He was reminded quietly on that occasion by B. B. Campbell that his recent service under S. C. T. Dodd made him vulnerable to the objection that he would be biased in favor of the Standard Oil, whose existence as a product of railroad favoritism the Cullom Committee had stressed emphatically.[11] Friends predicted that be-

[8] Sherman Papers, Sherman to Beebe, March 27, 1882.

[9] *Ibid., Pamphlets,* XX, No. 12, No. 18, No. 20, No. 21; Sherman, *op. cit.,* p. 75–76.

[10] Sherman Papers, Sherman to Whitman, November 17, 24, 1890, Sherman to G. A. Jenks, November 17, 1890, Sherman to Heydrick, December 2, 5, 1890.

[11] *Ibid.,* Campbell to Sherman, February 3, 1887.

fore many years elapsed Sherman would receive high judicial preferment.

During 1889 and 1890 he supported the Americanization program of the National Society of the Sons of the American Revolution. He became interested in historical research on George Washington's career, which he drew upon for an address to the Pennsylvania Society of the Sons of the American Revolution at Pittsburgh as its Vice President.[12] He joined the Society of American Civics, which encouraged instruction in government in schools and colleges. He joined the American Academy of Political and Social Science at Philadelphia, whose *Annals* provided data on current problems. All this stimulated his interest in governmental reform. He became an advocate of the Australian ballot.[13]

When the opera came to Titusville the Shermans sat "in the first row of the dress circle." Roger Sherman subscribed $500 in 1886 toward construction of the new Opera House.[14] There in late January 1889 he and Mrs. Sherman attended Jules Verne's "Michael Strogoff, the Courier of the Czar," whose ballets impressed the audience. The Shermans cultivated their interest in the dramatic arts further by participation in two associations whose activity enriched the city's cultural life. The most notable of these was the Shakespeare Club, which was patronized by a select group and celebrated the Stratford bard's birthday on April 23, 1883 at the Sherman residence. Roger Sherman's papers contain more frequent references to meetings of the XL Club on whose programs he presented essays and read parts of plays from 1886 to 1897. The XL Club flourished in competition with The East End Social Club, whose members read Shakespeare's plays during programs interspersed between gay parties.[15]

The Shermans' more limited means and Roger Sherman's stu-

[12] *Ibid.,* MS; Bates, *op. cit.,* p. 802.

[13] Bates, *op. cit.,* p. 802; Sherman Papers, receipts.

[14] Sherman Papers, receipts; Bates, *op. cit.,* pp. 362–363; *Titusville Morning Herald,* April 15, 1879.

[15] Sherman Papers, MS, *Pamphlets,* XIV; Interview with Miss Sarah Grumbine, February 9, 1958; Mrs. W. B. Roberts, MS Diary, 1881.

dious habits prevented indulgence in much of the gay social life of Titusville that characterized the wealthy circle. On the stormy evening of February 1, 1881, however, at a party for fifty at Mrs. David McKelvy's home Roger Sherman was the partner at whist of the cultivated, traveled, and fashionable Mrs. W. B. Roberts, the friend of Mrs. Vincent Astor. Mrs. Roberts belonged to the Shakespeare Club, with whose members she shared her gleanings from the art exhibits, opera, and concerts that she attended in New York City. Two years earlier Roger Sherman had attended the fiftieth birthday soiree held by the city's gentlemen for Colonel E. A. Roberts, inventor of the torpedo, and his associate in journalistic ventures.[16]

The years from 1879 to 1891 were years of quiet happiness in Roger Sherman's private life. On March 11, 1879, while he was heavily engaged in the Commonwealth and Ohlen litigations, his only son, Roger Seymour Sherman, was born. Happy and proud of his son, Roger Sherman was greatly troubled by Mrs. Sherman's subsequent severe illness. Prevented by this from attending General Sweitzer's hearings for nearly two months, he retained Dr. E. G. Cook of Buffalo as the family doctor. Dr. Cook's success in restoring Mrs. Sherman's health made him a warm family friend. Meanwhile, Roger Sherman reported happily to relatives and friends on young Roger's progress. He bought a horse and carriage for Mrs. Sherman. He acquired a cow from Beebe. He purchased a gray riding horse, which he rode regularly for exercise.[17] Sherman enjoyed fatherhood. As his son grew older, Sherman walked home from the office daily in time to romp with him in what Mrs. Sherman called "the usual after-tea circus." [18]

Almost a decade later he wrote *The Shermans,* his only book. Included in its combination of genealogies and personal reminis-

[16] Mrs W. B. Roberts, MS Diary, 1881, 1882–1885.

[17] Sherman Papers, Sherman to Dr. E. G. Cook, April 3, July 29, 1879, Sherman to S. Powell, August 1, 1879, Sherman to Cousin Fannie, December 5, 1879, Sherman to C. B. Seymour, April 14, 1879, Beebe to Sherman, January 6, 1881.

[18] *Ibid.,* Sherman to Beebe, January 20, 1881, Alma Sherman to Sherman, November 24, 1880.

cences is a statement of his personal faith: "Long descent is nothing; but the culture and growth of each individual in strength of mind and body is everything; fixed principles of citizenship of morals and of business conduct are everything; courage to assert and maintain conscientious and well-considered convictions is everything." [19] In 1880 he had confided to his Cousin Fannie his concern at the future effects upon the United States of the presence within it of the great mass of "ignorant negroes" and a larger number of "equally ignorant foreigners," and of an electorate "steeped in ignorance and corruption." "The standard of political morality is lower than ever before and intelligence is not obtained by bad education," he had observed gloomily.[20]

His personal integrity made him a highly respected counsellor for clients, a valued friend. The care with which he managed his wife's money inspired additional respect among informed citizens.[21] This respect was intensified by the support that he gave to the cause of state railroad regulation before the legislative committee investigating the desirability of a state railroad commission and by the draft bill that he prepared for it. Titusville's prosperity was injured by adverse rate and service discriminations.[22] The independent Oil Creek refiners would have benefited greatly from an equitable rate structure.

On June 4, 1892, an unprecedented catastrophe struck Titusville. Late in the day, after heavy rains had swollen Oil Creek, the Spartansburg dam upstream burst suddenly. The wall of high water suddenly flooded the Titusville railroad stations and the oil refiners' properties along the railroad tracks, and penetrated into the lower business district. The oil and benzine tanks of Rice & Robinson's refinery burst, their contents spreading on the flood and then igniting in a tremendous explosion at Schwartz's refinery below the city. The flames spread upstream to the International refinery, Rice &

[19] Sherman, *op. cit.,* p. 76.

[20] Sherman Papers, Sherman to Cousin Fannie, December 5, 1879.

[21] *Ibid.,* accounting with Alma C. Sherman, February 2, 1891, Sherman to James H. Caldwell, October 22, 1891.

[22] *Ibid.,* Sherman to S. J. Logan, February 21, 1891.

Robinson's, and the Oil Creek refining works, accompanied by explosions of stills and tanks. The sheet of flames burned all flooded buildings to the water's edge, including all the refineries, the lumber yards, the freight station of the Western New York and Pittsburgh Railroad and the Union Furniture Company. The Fire Department was helpless since the flood submerged the water works. Only the soaked roofs and walls of houses and business blocks saved the rest of the city from a general conflagration. Meanwhile, the heroic work of scores of men in small boats amid the flames rescued hundreds of stranded residents from factories and tenement houses. More than sixty died in the flood and fire. With several thousands homeless, scores of private homes were opened to provide them with shelter.

On June 5 at a City Hall meeting Roger Sherman was elected Chairman of the Voluntary Relief Association that was organized to cope with the catastrophe. An initial $3,000 was raised for relief. The *Morning Herald* reminded the city that it had contributed $15,000 to Chicago after its great fire. Sherman's committee immediately canvassed the city and nearby cities and towns for necessary funds. Governor Pattison on Sherman's request issued a proclamation appealing to the people of Pennsylvania to aid Titusville liberally. The Buffalo Express Company, the Western New York and Pittsburgh Railroad Company, and the Standard Oil Trust contributed. Philadelphia authorities declared the disaster "a second Johnstown." Philadelphians pledged $5,000 immediately and called a public meeting to raise more. Cleveland and Chicago followed this example. Local subscriptions rose to $18,000. A total of $104,000 was raised from 130 cities and towns in ten states and the District of Columbia.[23]

Sherman worked hard as Chairman of the Voluntary Relief Association. He served on the Committee of Finance which worked more than three weeks full time relieving the wants of sufferers after the Committee of Investigation examined the devastated areas

[23] *Ibid.,* Sherman to W. G. Hunt, June 6, 1892, Sherman to S. G. Decourson, June 6, 1892; *Titusville Morning Herald,* June 6, 7, 8, 22, October 31, 1892; Bates, *op. cit.,* pp. 453–457.

and studied the needs of property owners. Sherman received the Governor, State Board of Health, and committees from Philadelphia and Pittsburgh. He led in developing effective procedures and persuaded the Association to treat the contributed money as a trust fund. It then provided permanent relief somewhat after the method now pursued by the American Red Cross. Beginning with widows and their children, then continuing with families, the Association erected homes for them and found jobs for mechanics. Accounts were carefully kept. Beneficiaries of permanent relief were required to deed what remained of their properties to the Association. The Committee of Finance employed 400 men. In cooperation with the Board of Health and the city officials it repaired the bridges, restored the streets to passability, and re-established communications between all parts of the city. It dredged the channel of Oil Creek and then constructed a levee along it as a protection against future floods.[24]

Sherman's leadership did much to restore popular faith in the city's ability to recuperate from the disaster. His direction of the relief program received universal approbation. As the *Utica Globe* observed on July 2, the result of his labors "fully justified the high opinion entertained of his ability and worth." Mrs. Sherman served on the Committee on Lodgings which provided temporary housing for the homeless. Sherman contributed $100 initially and more subsequently to the relief fund, in addition to devoting full time for more than a month to his work as Chairman.[25]

From this service to Titusville he emerged as a civic leader of the first rank. He was a director of the Titusville Oil Exchange, a member of the Board of Trade that raised a large fund to attract industry and organized the Titusville Industrial Association to administer it. Sherman contributed $1,000 to this promotional fund. John Fertig, a director of the Association, and John L. McKinney conferred with him repeatedly about proposed industries. On Sep-

[24] *Titusville Morning Herald,* October 31, 1892; Sherman Papers, receipt from Voluntary Relief Association, Papers from E. O. Emerson, October 14, 1897.

[25] *Titusville Morning Herald,* June 7, 1892.

tember 27, 1896, he and Mrs. Sherman called upon Charles Horn, "the silk factory man," to persuade him to establish a factory in the city. Then Sherman drafted the formal contract that arranged this.[26]

Before making the popular pilgrimage to the World's Columbian Exposition at Chicago in 1893 Sherman supplied Henry Demarest Lloyd with printed records and briefs of the legal battles that he was fighting for the independent oil men. Lloyd was preparing his great classic, *Wealth Against Commonwealth.* Such material was invaluable grist for his literary mill. Sherman loaned him rare pamphlets on the petroleum industry. This contribution to Lloyd's collection of data surpassed that of any other person. In return the Winnetka antimonopolist loaned Sherman rare materials. To Lloyd's frank comments on the business situation that he was uncovering, Sherman replied with pungent candor. He confided to Lloyd details of the struggle to survive in which the independent Oil Creek refiners were engaged during the period when the price of refined oil was depressed before the United States Pipeline reached Wilkes-Barre. Sherman informed him of how he had defeated the injunction proceedings instituted against that novel pipeline.[27]

Sherman was gratified when Lloyd asked him in mid-May to read professionally the third draft of the manuscript of *Wealth Against Commonwealth.* Sherman promised to read the manuscript carefully, and said that while visiting "The Fair" in Chicago he would talk the matter over with Lloyd. "As for 'honorarium,' I doubt whether you will think me worth one, and if I think I have earned anything, I will leave it to you to say what," he added modestly. Lloyd sent the manuscript. Sherman read it, drafted his suggestions, and took it to Chicago. A short while earlier, after reading two-thirds of the manuscript Sherman had written to Lloyd, "I have no

[26] *Ibid.,* January 11, 1893, January 10, 1895; Sherman Papers, MS 1896 Diary; *Titusville Weekly Herald,* May 24, 1889, March 13, 20, 1896; *Titusville, Pennsylvania* (*Titusville Today*) (Titusville, 1896), pp. 1–8.

[27] Sherman Papers, Sherman to H. D. Lloyd, November 10, December 26, 1892, April 3, May 8, November 10, 1893, February 17, June 16, 1894; Lloyd Papers (Winnetka), Lloyd to Sherman, May 23, 1893, June 8, 1894.

doubt of its success." He added significantly: "As I read I could not avoid the thought that upon the subject you treat we have many ideas in harmony. *Have no fears of your book!* My feeble function of criticism died before I read many pages." [28]

Sherman and Lloyd met. The changes in *Wealth Against Commonwealth* that Sherman recommended protected it from any charge of libel. Lloyd invited the Shermans to The Wayside, his Winnetka home. Mrs. Sherman was unable to accept because of indisposition. Apparently Roger Sherman met there the charming, extraordinary Mrs. Lloyd. Like her husband she devoted her energies and income to aiding the underprivileged. At their table Sherman met other distinguished visitors to "The Fair" and probably some of the regular guests from the Chicago reform circle. For brilliance of conversation upon a broad range of topics, for brave analysis of the most difficult social problems of the decade, and for unflinching opposition to the business lords of the Trusts the Lloyd table and its hosts knew no peers in the middle west or on the Atlantic coast. Quiet, poorly clad denizens of the slums also sat at that round table of The Wayside, documenting by their presence the humanitarianism of the Lloyds.[29]

Sherman took Lloyd's manuscript back with him to Titusville, including a revised analysis of the position of the Interstate Commerce Commission. Sherman liked better "the scoring" Lloyd had given it at first, but remarked: "Perhaps it is more prudent and effective as you have it." He added revealingly:

I understand that the railroads threaten them from time to time with raising a constitutional position as to the legality of the Commission. And that some of them are afraid they may lose their places & salaries. My own idea is that the Commission has been carefully packed through

[28] Sherman Papers, Sherman to Lloyd, May 15, 25, 1893; Lloyd Papers (State Historical Society of Wisconsin), Lloyd to Sherman, May 23, 1893, Sherman to Lloyd, June 5, 1893.

[29] Sherman Papers, Sherman to Lloyd, July 20, 1893, Lloyd to Sherman, May 25, 1893; Chester McArthur Destler, *Henry Demarest Lloyd and the Empire of Reform* (Philadelphia, 1963), Ch. X and *passim.*

political & railroad influences. Whatever the cause may be it is completely emasculated, and might as well adjourn *sine die*.[30]

That bespoke the disillusionment of the Titusville and Oil Creek refiners who were still fighting to secure an equalization of the rates on oil in barrels with the rates on oil in tank cars. President Cleveland had appointed as the first Chairman of the Commission Judge Thomas Cooley, leader of the laissez-faire school of legal thought.

Shortly afterward Sherman returned Lloyd's manuscript. He also sent Mrs. Sherman's regrets that she had been unable to meet the Lloyds and her hope that she might be able to do so in the future. When Harper & Brothers required that Lloyd have his book read professionally for libel before publication Sherman did this. Later he verified the part of the galley proof that dealt with the petroleum industry. Despite this heavy encroachment upon his time he declined compensation, and supplied up-to-date data for insertion immediately before publication.[31] He declined to be quoted or to allow his name to appear in the text. However, without giving him credit Lloyd adapted and quoted his crushing description of the Pennsylvania Supreme Court's refusal of justice to the Producers Oil Company, Ltd., and the Producers and Refiners Oil Company, Ltd., in the Carter case.[32]

Sherman assisted Lloyd in the defense of his book. He informed him that the sharp attack upon it in the New York *Nation* had been written by W. T. Scheide. When George Gunton attacked Lloyd's book in the *Social Economist* in "The Integrity of Economic Literature," Sherman informed Lloyd that Gunton was a secretly paid Standard Oil propagandist. This the Rev. Washington Gladden verified. Lloyd drafted a reply, which Sherman read and approved in

[30] Sherman Papers, Sherman to Lloyd, July 20, 1893.

[31] *Ibid.*, July 20, 29, November 10, 1893, February 17, 21, April 13, 30, May 3, 15, 30, June 2, 8, 16, 1894.

[32] Lloyd Papers (Winnetka), Sherman to Lloyd, June 8, 1894 with MS alterations in Lloyd's hand.

advance of publication. When this appeared in the *Boston Herald* Sherman distributed copies in the Oil Region.[33]

When the Standard Oil's prestige cracked under the impact of *Wealth Against Commonwealth* John D. Rockefeller sought exoneration from the leading Protestant evangelists, Social Gospel clergy, and new school economists. Washington Gladden and Lyman Abbott, editor of *The Outlook,* informed Lloyd of these invitations to meet with Rockefeller and the Rev. B. Fay Mills at 26 Broadway to hear the Standard Oil's defense. The invitations stipulated that Lloyd was to be absent while his book was rebutted. Lloyd informed Sherman of the coming event. The Titusvillian was vastly amused by Rockefeller's attempt to secure exoneration from "a committee of *preachers!*" and called it "one of the roaring farces of the day." Lloyd's analysis of the significance of the invitations, when sent to the invitees, so briefed them that they refused to attend the "investigation" unless Lloyd were present and allowed to reply. This precipitated the abrupt abandonment of Rockefeller's attempted coup.[34]

Sherman was deeply moved by *Wealth Against Commonwealth.* He wrote Lloyd that he had performed "a most noble and effective work." He continued: "Your book should be the 'Uncle Tom's Cabin' of this Era, and I pray that it may be. Your summing up in the chapters in conclusion and that upon the 'Smokeless Rebate' cannot fail to be as effective as it is great and unanswerable." The questions Lloyd raised "are of the most momentous importance. How I would like to join in the coming crusade for Liberty . . . what Titan efforts must be made to save popular gov-

[33] Sherman Papers, Sherman to Lloyd, January 11, September 16, October 15, 29, December 30, 1895; *New York Evening Post,* November 10, 1894, reproducing the *Nation* review; *Social Economist,* IX (New York, June 1895), 11–25; *Boston Herald,* October 23, 1895.

[34] Sherman Papers, MS 1896 Diary, May 2, 1896, Lloyd to Sherman, May 22, 1896; Lloyd Papers (State Historical Society of Wisconsin), L. Abbott to Lloyd, May 19, 1896, Lloyd to B. F. Mills, May 12, 21, 1896, Lloyd to R. T. Ely and enclosures, April 29, May 22, 1896; Lloyd Papers (Winnetka), Lloyd to Gladden, April 23, 1896, Mills to the invitees, April 22, 1896 (copy).

ernment and true liberty!"[35] William C. Plumb agreed in the *Bradford Daily Record.* In no other book, he declared, "is there such conclusive evidence of the power of plunder to buy out the law." The Titusville *Sunday World* recommended that *Wealth Against Commonwealth* be used as a textbook in colleges and schools. If acted upon by the independent oil men and proved in court, the facts it contained, it declared, "would send all the leaders of the Standard monopoly, together with many a railway manager, to the penitentiary."[36] Sherman had informed the independent oil men and press of the Oil Region that the book might help their cause.

Lewis Emery, Jr., personally distributed more than 100 copies to influential Americans. A. D. Wood, Secretary and Treasurer of the Producers Oil Company, Ltd., commended the book's "masterly" arraignment of the Standard Oil's "monstrous, criminal monopoly," and voiced the independents' bitterness at the oil monopoly's ruthless and corrupting methods and at the government's failure to protect their right to engage in business. "Our independent people have recommended and circulated your book as widely as possible," he informed Lloyd, and thanked him for his "championship."[37] Sherman had not forgotten his responsibility as the chief counsel of the United States Pipe Lines Company in assisting Lloyd. As Lloyd's chief voluntary assistant and legal counsel, and as a leading promoter of the circulation of *Wealth Against Commonwealth,* he contributed to that attempt to revolutionize public opinion and overthrow the entire system of "Bourbon" business and political power.

Sherman also was developing an intellectual approach to the problems of the "Mauve Decade." His Washington's Birthday address of 1895 to the Sons of the American Revolution at Pittsburgh

[35] Lloyd Papers (State Historical Society of Wisconsin), Sherman to Lloyd, October 28, 1894.

[36] *Bradford Daily Record,* November 26, 1894; Titusville *Sunday World,* October 28, November 17, December 29, 1894, February 19, 23, 1895.

[37] Lloyd Papers (State Historical Society of Wisconsin), Emery to Lloyd, May 23, 1896, Wood to Lloyd, March 19, 1895.

expressed his viewpoint cogently. This expressed his admiration for Washington and loyalty to the American faith while criticizing the defects of the nation. The typical ideals of Americanism which the Society should safeguard he found in *noblesse oblige* and in "liberty subjected to law." The latter should restrain "liberty that dares to do wrong to the equal" and "law subjected to liberty." Growing lawlessness was a great national danger, he declared bluntly.

> Lawlessness in high places; lawless business methods; lawlessness of public men; . . . The constant spectacle of legislators faithless to their obligations to their constituents and to the State; of corrupt politicians escaping punishment, and holding places once considered honorable, by grace of a dollar; of great corporations and combinations of capital, lifting themselves beyond the reach of the individual citizen, and, in some instances, beyond that of the Commonwealth itself, can but breed other lawlessness. . . .

Hence, the true American should "hold all citizens to obedience to law." This Americans could not do if they played the role of "a feeble race of men." Daniel Webster was his authority for insinuating that the concentration of wealth was contrary to the American way. He warned against the popular belief that Americans were exempt from the calamities that befell others. He criticized barbarous election methods. The true patriot "must choose the ground upon which he will stand to fight again a battle for the race." This "revolution" would advance humanity, popular government, and racial progress, and maintain law and order among "a self-governed people, freed from industrial tyranny and the domination of the Golden Calf." For this brave stand Lloyd wrote Sherman glowing praise.[38]

More philosophical was Sherman's anonymous editorial, "Mistakes of the Plutocrat," in the Titusville *Sunday World,* July 27, 1895. This castigated the avarice and social Darwinism that justified to the rich the deprivation of opportunity for the poor to rise to higher levels. The wealthy were blind, he declared, to everything

[38] Sherman Papers, MS "Address" February 22, 1895, Lloyd to Sherman, March 5, 1895.

that was not "business" and included in "the one idea of money-getting." These were mental and moral defects together with "insolence and arrogance" that accompanied great wealth acquired by wrong methods. This explained the plutocrats' hatred of reformers who were seeking to improve "the conduct of human affairs." This attitude derived from fear that the plutocrats' critics were jealous of their wealth and desired to redistribute it more equitably. It was not the wealth of Rockefeller or Gould that made them objects of reprobation. "It is the methods they employ in getting the money, and the bad uses they make of it." In the Oil Region "the most offensive development" was seen in the Standard Oil Trust's lashing itself into "fury and falsehood" over Lloyd's *Wealth Against Commonwealth* and F. F. Murray's *Middle Ten.* The latter urged that the great middle class could rule America "if it knew its power" in opposition to both the anarchists and "that class equally destructive of civilization, whose god is money."

Sherman was impressed by Murray's book. It urged the middle class to free itself from the professional politicians and their deceptions, corruption, and support of Standard Oil business oppression. Their activity, Sherman declared, was leading to business depression "in many localities," the turbulent uprising of labor, and "the division of society into two classes—the wealthy employer and the poverty-stricken employee." Thus he articulated the emergent theory of all critics of the regime of "Big Business." [39]

Additional insight into Sherman's thought can be gleaned from a program contribution to the XL Club, a married-couples club. As President, on February 11, 1895, he included many women members in the "President's Programme." They criticized the education of women and discussed public and private ethics and the effect of political corruption on general morality. Sherman's part of the program included a quotation from the Rev. Josiah Strong, the conservative Social Gospel leader, which predicted a third great era of discontent analogous to the Reformation and the French Revolu-

[39] Sherman Papers, MS "Mistakes of the Plutocrat"; Titusville *Sunday World,* July 27, 1895.

tion, which would terminate "economic despotism." Significantly, he also read some verses attributed erroneously to Alcaeus:

> What Constitutes a State. . . .
> High minded men
> With powers as far above dull brutes endured. . . .
> Men who their duties know,
> But know their rights, and knowing dare maintain
> Prevent the long-aimed blow,
> And crush the tyrant while they rend the chain.[40]

[40] Sherman Papers, MS "Quotations," February 11, 1895.

# CHAPTER XII

## *Again Counsel of the Independents*

SHERMAN did not regain immediately his former position as chief legal adviser of the independent oil men after termination of his Standard Oil contract. Undoubtedly his decision to return to private practice paved the way for this, as did his editorship of the *American Citizen.* His robust antimonopolism had never been more trenchant than in his editorial of March 11, "What is a Monopoly?" In December he attacked the Standard Oil's advantages on the railroads. Such championship appealed to the reviving competitive spirit of the Oil Region. Sherman was at peace again with himself as he renewed combat with the Rockefellers.

Early in 1887, as his service with the Standard Oil terminated, Sherman joined John Fertig and other Titusville producers in organizing the National Oil Company. That firm drilled in Allegheny County and subsequently in the Lima field in northwestern Ohio. In 1890 Sherman became active in the management of the Orion Oil Company, Ltd., which was also organized by Titusville men. After preparing this second company's articles of association Sherman subscribed to one-eighth interest of the $40,000 capital. With this small fund the company successfully drilled paying wells and then sold them quickly for a profit. Its operations were on the western border of the now-declining Oil Region. Sherman was again a small producer. His capital investment in oil production could not have been more than $15,000. His profits were undoubtedly moderate. He was now in a position to join in a new producers' movement and to renew his friendship with Lewis Emery, Jr.

The hostility of the Pennsylvania independent producers and

refiners to the Standard Oil had flared up again in another bitter protest in February 1887 just as Sherman was beginning to enjoy professional freedom. The explosion was precipitated as had been that of 1878 by the decline of crude-oil prices to a record low level, but its force derived from a mounting list of genuine grievances. Of their reality Sherman was all too cognizant. The protest set in train a course of events that drew him inexorably into the movement to re-establish competition in the petroleum industry in Pennsylvania.

The producers had instituted, from 1884 to 1885, a shut-down movement after ineffectual negotiations with the Standard Oil. However, the managing committee had failed to persuade the producers of the Thorn Creek field to shut down their gushers. This had rendered the widespread shut-down elsewhere ineffective. The crude-oil price fell to 73¢ a barrel on January 6, 1885, when the shut-down was discontinued. Competitive drilling in new fields in Butler County and in the widening Bradford field had increased oil in storage to 33,395,885 barrels by January 1887, and pushed the price of crude oil down to 54¢ by the following July.[1]

At these low prices the service rates of the Standard Oil pipeline subsidiaries became an intolerable burden, whether the 20¢ per barrel charge for gathering oil at the wells or the 40¢ per 1,000 barrels per day storage charge plus "shrinkage" of 3% per year. Except for those fortunately possessing gushers or other large producing wells these charges were exhorbitant.[2] Although overproduction was technically the basic cause of the producers' dilemma, they knew that as the chief purchaser the Standard Oil was the chief beneficiary of abnormally low crude-oil prices. Many producers knew that the market price of refined-oil products had not been reduced proportionately. They believed that pipeline charges should be reduced so as to enable them to continue in business. Furthermore, it was widely known in the Oil Region that the Standard Oil still enjoyed secret railroad rate and service favors. This had been

[1] *Petroleum Age,* III (June 1884, January 1885), 755–756, 887–889; Johnson, *op. cit.,* p. 130.

[2] Johnson, *op. cit.,* p. 150.

disclosed by successful lawsuits that Ohio independents had instituted against the railroads.[3]

In 1886 the independent producers had contributed to the election of Governor Robert E. Pattison on an antimonopolist platform. Pattison demanded that the new legislature enact statutes to enforce Article XVII of the Pennsylvania Constitution forbidding transportation discriminations. "Discrimination in charges and facilities for transportation is as widespread and injurious as it ever was," he declared.[4]

In January 1887 Representative James K. Billingsley of Washington County introduced an ill-conceived bill into the legislature which was designed "to punish pipelines," companies, and persons engaged in the petroleum pipeline business if they violated the regulations of rates and services that it provided. Reasonable lower rates for piping and storage were specified. In this bill, for the first time, the producers evinced an appreciation in the legislature of the key contribution that pipelines made to the Standard Oil's dominant position in the industry. Thus, while the Interstate Commerce bill was on its way to final enactment by Congress, the producers demanded that Pennsylvania regulate the pipelines more effectively.

Unfortunately, the independents ignored the constitutional prohibition against punitive laws. Also, the Billingsley Bill attempted regulation by statute without having first ascertained carefully the facts and conditions of the industry. The bill's purpose was to reduce pipeline charges to a reasonable level and to prevent service discriminations. The fact that much of the pipeline business was interstate and beyond Pennsylvania's jurisdiction escaped notice. Some of the proposed regulations were onerous and unworkable.[5]

The Billingsley Bill was supported on February 14 in the *Pittsburgh Daily Dispatch* by T. W. Phillips of Butler, a large producer, who reviewed the power, misdeeds, monopolistic practices, and

[3] Nevins, *Study in Power,* II, Chs. XXIII–XXIV; Tarbell, *op. cit.,* II, Chs. X–XI.

[4] Johnson, *op. cit.,* p. 130, quoting *Pittsburgh Dispatch,* January 5, 1887.

[5] See Johnson, *op. cit.,* pp. 130–131 for details.

political methods of the Standard Oil in a vivid analysis.[6] The producers lobbied at Harrisburg for the bill's enactment. In addition to much-needed relief to them it offered a long-desired means of punishing Rockefeller and his monopoly. S. C. T. Dodd, the Standard Oil's General Solicitor, immediately counterattacked with the aid of its subsidized *Oil City Derrick, Bradford Era,* and *Titusville Morning Herald.* A Republican legislator, Wallace Delemater of Crawford County, organized opposition in the legislature to the bill while Dodd riddled it in legislative hearings.

In successive *American Citizen* editorials Roger Sherman defended the principle upon which the bill rested. The petroleum pipelines, he asserted, were public utilities in law and their charges could be regulated. While observing that many Oil Region citizens were telegraphing protests and seeking the measure's recommitment he observed that Butler County, Pennsylvania, and Olean County, New York, had most of the pipelines. The bill would necessitate federal regulation of pipeline charges, he added, an oblique method of indicating that it really should be introduced into Congress. While commenting upon the bitter name-calling that the bill had precipitated between the Standard Oil and Lewis Emery, Jr., and David Kirk, who were the bill's leading legislative advocates, Sherman supported the Titusville committee that was seeking workable amendments. This committee urged also that the producers and the National Transit Company, which now operated all Standard Oil pipelines, negotiate an agreement that would remedy existing ills. In reply to the National Transit Company's protest that the bill would impair the obligations of contracts. Sherman observed that this would not apply to oil received for piping after its enactment.

The great principle embodied in the Billingsley Bill, he asserted, was the state's right to regulate and control. "This should be established in Pennsylvania as it has been in other states," he declared. The issue was intimately related to the recent growth of monopolies

[6] Reprinted in *Petroleum Age,* VI (April 1887), 1597.

and the public's conclusion that they were evil. "Whenever a practical monopoly has become established," Sherman ruled, "it must be controlled." [7] While he obviously did not approve of the measure's unworkable sections, Sherman applied to industrial monopoly the theory of the regulatory state which he had helped to formulate earlier in the fields of management-labor relations, pipelines, and railroad transportation. Thus, while meetings were held in the principal cities of the Oil Region to discuss the bill, Sherman attempted to hold the discussion to the central issue.

The *Petroleum Age* observed that the corollary purpose of the Billingsley Bill was also to prevent the National Transit Company from destroying the newly established independent refineries as the railroads had done to their predecessors in the Standard Oil's behalf. This statement explained the reason behind the leadership of State Senator Lewis Emery, Jr., in the fight for the bill. While continuing to be a successful producer he had recently established a refinery with his partners near Philadelphia, only to discover that he was discriminated against in the joint railroad-pipeline crude-oil rate. He and his partners, the firm of Logan, Emery and Weaver, retained Roger Sherman as counsel and instituted suit against the Pennsylvania Railroad to recover excess charges of $50,000. Thus Emery and Sherman again joined forces in behalf of the independent oil men's reviving attempt to overthrow the Standard Oil monopoly. The Emery brothers, David and Lewis, regarded Sherman as the smartest lawyer in Pennsylvania.[8]

The defeat of the Billingsley Bill in the Pennsylvania Senate by a vote of 25 to 18, with seven Senators absent or not voting, was "a crushing blow to the great number of oil producers who saw in it an attempt to relieve them from a few of the burdens under which they have been laboring." The "contemptible treatment" accorded Emery

[7] *American Citizen,* February 18, March 18, 25, 1887.

[8] *Petroleum Age,* VI (February 1887), 1551, (March 1887), 1573; Sherman Papers, A. H. Logan to Emery, Jr., March 31, 1887; Johnson, *op. cit.,* pp. 134–135; Interview with Mrs. Lena Emery Brenneman, June 22, 1942.

when it was called up for a vote on April 28 proved to them "conclusively that the oil country had nothing to expect from the legislators," the *Petroleum Age* observed.[9]

Immediately after the bill's rejection the producers' delegations met at the Leland House in Harrisburg. The meeting adopted resolutions recommending, after this defeat of their "final appeal asking only a reasonable restriction of this tyrant power," that the producers organize and set up immediately a "cooperative joint stock company" embracing "the people of the entire oil fields." Committees were to be appointed for each district of the Oil Region, which were to send delegates to a general meeting at Oil City, while a drafting committee headed by T. W. Phillips was to produce a plan of organization. The Oil City meeting produced the Petroleum Producers' Protective Association, headed by Phillips, with thirty-six secret local assemblies and 2,000 members.[10]

Meanwhile, in preparation for trial of Logan, Emery and Weaver *v.* Pennsylvania Railroad Company, Lewis Emery, Jr., sent Sherman confidential data that he had secured from John Teagle of the independent oil-refining company of Scofield, Shurmer and Teagle of Cleveland, Ohio. This data disclosed rebate discriminations against Logan, Emery and Weaver and the Oil Creek refiners. After this suit went to trial and proceeded for three days the Pennsylvania Railroad offered to settle out of court for $30,000 after the precedent set by the Ohlen suits. Sherman accepted this in behalf of his clients because the court had applied to this civil suit the criminal law principle prohibiting self-incrimination so as to deny the plaintiff an order obliging the railroad to produce the written statements, vouchers, and cancelled checks that would establish after June 3, 1883, the facts of rebates to the Standard Oil and overcharging of Logan, Emery and Weaver. Sherman understood that the court was prepared to deny its jurisdiction on the ground that the oil shipments which had passed from northwest Pennsylvania through a

[9] *Petroleum Age,* VI (May 1887), 1633; cf. Williamson and Daum, *op. cit.,* pp. 558–562, for the best account of the Billingsley Bill.

[10] *Petroleum Age,* VI (May 1887), 1633; Johnson, *op. cit.,* pp. 136–137; Tarbell, *op. cit.,* II, 158–159.

corner of New York State to Philadelphia were interstate in character. Logan, Emery and Weaver had gone to court not only to recover the overcharges but also to secure a remedy for the discriminatory rate. Unable to win the latter, the firm sold its refinery, which had been an important market for independent oil, to the Standard Oil, and discontinued certain suits against it.[11]

Sherman's professional and personal relation to Senator Emery continued to be friendly. Emery assured the *Oil News* and Henry Demarest Lloyd that despite the sale he continued to be an independent.[12]

Lewis Emery, Jr., was active in the Petroleum Producers' Protective Association, and Sherman supported it warmly in the *American Citizen.* When Phillips proposed a shut-down movement Sherman declared in that paper that the attempt to restrict production so as to restore prices to a paying level was entitled to the sympathy of "moral sentiment" since "wild speculation" on the exchanges was partly responsible for the producers' plight.[13] Despite the bitterness of the producers against the Rockefeller organization, with the aid of John L. McKinney of Titusville and the astute Warren attorney, J. W. Lee, Phillips persuaded the Association to join with certain large producers and accept Rockefeller's offer to reserve the speculative profits on 6,000,000 barrels of oil for producers and members of the oil-well-workers' union cooperating in the shut-down.

The incentive provided by this prospective profit led to general producers' participation. Production was reduced to between 10,000 and 15,000 barrels daily below market demand. Surplus oil stocks were reduced by 15,000,000 barrels during the shut-down, the *Morning Herald* claimed. After such an achievement Sherman urged in the *American Citizen* that the producers retain "their

[11] Sherman Papers, Sherman to Lloyd, January 16, 1892, *Paper Books, Appeal of the Pennsylvania Railroad Company in Logan, Emery and Weaver v. Pennsylvania R. R. Co., Supreme Court of Pennsylvania, Eastern District, No. 123, January Term, 1890.*

[12] *Ibid.,* Lloyd to Emery, Jr., December 14, 1887, Sherman to Lloyd, January 16, 1892; Hidy, *op. cit.,* p. 214.

[13] *American Citizen,* December 9, 16, 23, 1887.

present magnificent organization."[14] The price of oil rose from a low of 43¢ a barrel to more than 91¢ in June 1889, when the shut-down terminated with the sale of the last of the 6,000,000 barrels and the division of the profit. The arrangement saved many producers from bankruptcy, which was to the Standard Oil's undoubted advantage. In retrospect, as the *New York Tribune* observed, the cooperative shut-down had been "one of the most interesting economic experiments" of recent years.[15] When the booming market for petroleum products, combined with the improved storage situation, pushed Pennsylvania crude oil up to $1.01 on June 25, 1889,[16] the price phase of the producers' complex dilemma was greatly alleviated.

Meanwhile, the independent oil men had capitalized upon unusual opportunities to publicize their opposition to the Standard Oil's unfair competitive methods. Sherman vigorously backed the *New York World*'s sensational protest against the wrecking tactics of a Standard Oil subsidiary, the Vacuum Oil Company, that had been exposed in an extraordinary Buffalo conspiracy trial.[17] George Rice, the independent Marietta, Ohio, refiner, continued his effective pamphlet exposures of Standard Oil marketing methods, interspersed with damage suits against the Standard Oil Trust when he found legal openings in its armor.

Although a parallel New York Senate investigation proved to be a farce,[18] in 1888 a congressional investigation of certain trusts paid effective attention to the Standard Oil. Franklin B. Gowen, John Teagle, Lewis Emery, Jr., Rice, Charles B. Matthews, and other

[14] November 30, 1888.

[15] *New York Tribune,* quoted in *Titusville Morning Herald,* July 3, 1889; *Titusville Morning Herald,* May 16, 1889; *Derrick's Handbook,* pp. 470–472; *Investigation of Certain Trusts,* pp. 60–61; Johnson, *op. cit.,* pp. 138–139, 168–169; Williamson and Daum, *op. cit.,* pp. 563–568; Tarbell, *op. cit.,* II, 159–162. Later authorities assert that the reduction in oil stocks ranged from 8,600,000 to 10,000,000 barrels.

[16] *Titusville Morning Herald,* July 26, 1889.

[17] *American Citizen,* May 20, 1887, *et seq.;* for the record of that case see Erie County (N. Y.) Clerk Archives, The People of the State of New York *v.* Hiram B. Everest and another, *Bill of Exceptions.*

[18] *Toledo Blade,* February 28, 1888, quoting *New York World.*

spokesmen of the independent oil refiners and producers, and E. F. Howes, manufacturer of oil-well supplies, described for Congress the Rockefeller organization's business espionage, its driving of crushed refiners into oil production which had increased output beyond demand, and the resulting reduction of drillers' wages as the crude-oil price fell below the cost of production. Paralleling this was evidence adduced to show that the earlier railroad discriminations favoring the Standard Oil were continuing. The validity of all this Archbold denied, while Dodd inserted into the record of the hearings an expurgated record of the Buffalo conspiracy trial.[19]

Most of this anti-Standard Oil testimony fed the swiftly rising antitrust sentiment of the nation. Gowen declared that Standard Oil control of the pipelines demanded remedial action. On April 28, 1887, after defeat of the Billingsley Bill, the defeated producers had proposed producer-owned pipelines to extricate themselves from their dilemma.[20] Gowen proposed to the congressional Committee on Manufactures that the states expropriate the petroleum pipelines, that the Interstate Commerce Act be amended to give the Interstate Commerce Commission authority over interstate pipeline operations, and that federal mandamus proceedings against pipeline companies be authorized.[21]

Gowen's proposals were provoked undoubtedly by the growth of the Standard Oil minority interest in the Tide Water Pipe Company after it had been obliged to enter into a pool with the National Transit Company early in the decade.[22] The testimony of State Senators Emery and Kirk during the congressional investigation made it evident that the Standard Oil pipelines were overcharging producers and other independent oil shippers grossly.[23] Those pipelines were linked with the railroads in petroleum traffic pools that elevated joint rates to exorbitant figures.[24] Gowen's constructive proposals were fruitless. It was obvious that Standard Oil-controlled

[19] *Investigation of Certain Trusts, passim;* Schlegel, *op. cit.*, pp. 279–285.
[20] Johnson, *op. cit.*, p. 138.
[21] *Argument of Mr. Franklin B. Gowen*, p. 24.
[22] Tarbell, *op. cit.*, II, 19–23; Hidy, *op. cit.*, pp. 220, 609.
[23] Johnson, *op. cit.*, pp. 151–152.
[24] *Ibid.*, p. 150.

pipelines were the major obstacle between producers and the market.

Roger Sherman had quietly briefed Senator Emery, his client, for his testimony before the congressional Committee on Manufactures. Now Sherman became involved quickly in the attempt of leading producers and the independent refiners of the Oil Region to free themselves permanently from dependence upon the National Transit Company's gathering and long-distance crude-oil pipeline system and also from continuing railroad favoritism to the Standard Oil in oil-products transportation. They perceived correctly that the Trust's pipelines were the chief bulwark of its supremacy in refining and a major source of its profits.[25]

Early in 1889 Sherman joined Emery, Phillips, Kirk, Lee, and other leaders of the producers in secret conferences on the problem of crude-oil pipeline transportation. Sherman's earlier association with the producers' Seaboard Pipe Line Company and the Tide Water Pipe Company made him especially valuable as a consultant on the legal problems and larger aspects of the business situation that were involved in the revival of the attempt to contruct and operate an independent crude-oil pipeline to the seaboard. Eventually he was asked to draft the plan of organization. This suggests that he had again assured the independent producers and refiners that they could finance and would profit from an independent gathering pipeline system feeding an independent trunk pipeline to the Atlantic coast. All the conferees were dissatisfied with the dilatory proceedings before the Interstate Commerce Commission, where the independent refiners' associations of Titusville and of Oil City were challenging the pooled rates on refined oil which had resulted from the railroads' agreement with the Standard Oil. Sherman had withdrawn from this proceeding abruptly. Other hearings had been initiated on the practice of charging higher rates on oil in barrels than on that carried in tank cars.[26]

[25] *Ibid.*, pp. 151–152, 273 n.63.

[26] Sherman Papers, R. Reynolds to Sherman, February 1, 1889; Interstate Commerce Commission, *Third Annual Report* (Washington, 1890), 155–156; Titusville *Sunday World*, May 19, 1894.

In September 1888 the railroads had increased barrel rates from Titusville to Perth Amboy some 27%, which made it impossible to refine oil in the Oil Creek area and compete in New York with the refined product of the seaboard. Gowen replaced Sherman at the Titusville hearings and attacked the pooling agreement between the Pennsylvania Railroad and the National Transit Company as discriminatory. He also brought up the Standard Oil's refusal to produce this document lest doing so incriminate it. The testimony strengthened the independents' conviction that the profits of pipeline transportation gave the Standard Oil Trust an almost decisive advantage. The new independent refineries that had sprung up in the Oil Region under the encouragement of the Interstate Commerce Act desired to tap the export market. Lacking tank cars they wished to reach this with barreled illuminating oil and other products.[27]

The Petroleum Producers' Protective Association was almost demoralized when its members learned in June 1890 that they had been nearly taken into camp by H. L. Taylor of the Union Oil Company. That firm's ownership by the Standard Oil since 1883 had been kept secret. Taylor had joined the Association and proposed that his company be made the instrument of the independent producers' cooperative marketing of their oil. Taylor had been made chairman of a committee to explore such a policy before he was forced to admit that he and his company *were* Standard Oil. Whereupon his resignation was demanded, secured, and he was ousted from the Association abruptly. When Rockefeller began distributing his Trust's certificates to his producing subsidiaries (after Phillips had sold a large producing property to the Standard Oil) a stampede of independent producers to his standard was prevented only by strenuous efforts of the Association's leaders. Emery, Lee, Kirk, A. D. Wood, Rufus Scott, and Sherman arranged for a meeting of the Association's General Assembly at Warren on January 28, 1891. During a three-day session there the Assembly abandoned

[27] *Titusville Morning Herald,* May 16, 1889; Johnson, *op. cit.,* pp. 144–145.

Taylor's cooperative plan. Instead, it adopted the policy of withholding the independent producers' oil from the Trust and of organizing limited partnership associations to handle this, marketing it either to the independent refineries or to others to be built with producer-supplied capital.[28]

Sherman was on guard against diversions of the energy and force of the Petroleum Producers' Protective Association into lesser and unprofitable channels. Among these was a renewed attempt to secure state pipeline regulation in Pennsylvania by means of the Burdick Bill of 1891.[29] Sherman's caution rested upon a realistic appreciation of the weakness of earlier producers' movements and of the chief motive inspiring them. As an investor in two oil-producing companies financed by Titusville men, he was cognizant of the problems of the producers. He knew also how the Standard Oil misrepresented and attacked the high-caliber leadership that his friend and client Senator Emery gave to the independents and how indispensable he was to the Association.

He warned Emery on February 20, 1981, that the Association's energies must be concentrated upon a more fruitful and appealing course and against ill-advised personal action in refining that would betray the producers' faith in his leadership. This letter illustrates the business acumen that Sherman had developed during his corporation practice and which he now focused upon the strategy of the independent movement:

> As you know, and everybody knows, I am in favor of the passage of the "Burdick" Pipe Line bill, upon principle and broad grounds of public policy. If passed its influence as to a precedent will be of great value in relation to other needed legislation. It will be particularly beneficial to the Oil trade, in my opinion, because it will tend to take the business out of the rut in which it has been rolling along, and out of the revolution which it will create, we may hope for better conditions. The *direct* benefit which the oil producer will get from it will be small.
>
> Having said this much, I desire to say more, and it must all be

[28] Tarbell, *op. cit.,* II, 162–165.

[29] Cf. Johnson, *op. cit.,* pp. 164–170.

understood and taken in connection with and positive affirmation of what has been just stated.

All previous movements of and by the producers have been stranded and wrecked upon one or both of two rocks, the rock of "buncomb" and that of private, personal and selfish interest. They have weakened themselves, always, by mass meetings, flaws, & high sounding resolutions; there has been a lack of what is called "business," by which I understand that faculty which divests itself for the time being of sentiment, passion, or revenge, and looks solely and coldly at the practical results to be obtained by proper organization, & by the steady and persistent following out of well considered plans and methods, with a view to protect property rights and the acquisition of pecuniary profits which should naturally result from energy, enterprise, intelligence, labor and capital. I do not believe it wise for the leaders in the present producers' movement to spend all of their energies and time, nor any great part of either in trying to make the Burdick bill a law, to the exclusion of other things of equal if not of greater importance. I believe that the moral effect is bad, and that by such a course you will alienate many men whose support is most desirable. I do not speak unadvisedly. I have talked with many, and I find a feeling of distrust and of dissatisfaction springing up, and an impatience with inaction in more promising fields. I do not believe this bill, or any other bill, if they were now laws would be curealls for the diseases of the producers' business. The problem of what to do and how to do it would still remain to be solved, and their solution is possible only upon the lines that I have indicated.

It seems to me that at the earliest possible moment a full meeting of the Executive Board should be had, and that pending the proposed increase of its numbers to eleven at least two representative men from the Pittsburgh field should be invited to attend, and their cooperation secured. There should be a general discussion of ways and means, including proposed legislation, and a plan or plans devised by which that greatest lever in these days, and possibly in all other eras,—self interest—can be brought into play to move the oil producer to think about his salvation.

Do not understand me as criticising or finding fault with what has been done. It is all well enough; but the time has come when something more should be done, and done speedily.

We have an excellent organization, good men, an excellent spirit prevailing, and all that is needed is concentration of thought and action upon a practical money-making plan. . . .

P. S. The enemy is persistently circulating the report that you are about to build, or capitalize, or become interested in, a refinery in Philadelphia or New York, and that you have a personal interest in the Burdick bill. It is said here, and I heard it in Warren Wednesday, that you propose to start your son in the refining busines at the Seaboard. I hope there is nothing in it, and that you will promptly and publicly deny the story. I hope further that you will not do any such thing at present if ever. For any member of the Executive Board to have a personal & separate interest, apart from the general mass of producers whom he represents and in any contemplated action of the Board, would be a death blow to the movement.[30]

On the following day Sherman wrote firmly in a similar vein to the influential J. W. Lee that the producers' movement should concentrate its energy in the business field rather than waste it in another futile legislative foray. The press reported that the House committee had "agreed to report the Burdick Bill with a negative recommendation." He regarded it "as very unfortunate that the producers have antagonized so many who should be their friends at the beginning of the present movement." There were, he added, "very many good men, whose sympathies are with us, and who would gladly unite on a business proposition, who fear that the present organization will go to pieces in efforts to impress the unimpressionable, and in public meetings and legislative efforts which can produce no immediate practical results." Several business men of this type of pecuniary means and talent had expressed themselves to this effect to him. While the Burdick Bill was right in principle and should not be abandoned, Sherman said, "the time has come when prompt and decided action should be taken in other directions. It seems to me that there should be an early and full meeting of the Ex. Board and a full discussion of the situation." [31]

[30] Sherman Papers.
[31] *Ibid.,* Sherman to J. W. Lee, February 21, 1891.

As an influential leader of the Petroleum Producers' Protective Association Sherman was pressing for action on the pipeline project and allied business ventures.

He had learned much since 1878. The contrast between the position taken in these letters and the fourfold strategy of the producers of 1878 to 1800 is striking. The defeat of the Billingsley Bill was very much on his mind. The proper field in which to strike at the Standard Oil was in business, its greatest citadel, he clearly implied. After all was said, Benson's success with the Tide Water, partial and short-lived as it had been in reviving competition, had been the only substantial achievement of that second great upsurge of the producers. In politics the Standard Oil and the Pennsylvania Railroad could always rally the decisive support of the Cameron Republican organization. This Sherman appreciated as the Democratic leader of northwestern Pennsylvania perhaps more fully than did Emery, who was a Republican.

In his stand Sherman was entirely successful, in part because of his influence upon Emery. The Executive Board of the Association met, canvassed the situation, and drafted a statement of policy. This proposed to organize the Producers Oil Company, Ltd., as a producers' cooperative agency with a capital of not less than $500,000, subscribers to which should be members of the Association in good standing. This proposal was circulated to the membership, in part by Lee and J. R. Goldsborough, Secretary of the Association, as an agreement setting up a schedule of ten payments of each subscription, to be filled in with the amount of the subscription and signed. As this document stated, the objective of the Producers Oil Company, Ltd., was "to obtain, for the producers of petroleum, a competitive market for their product, by securing to them the control of the oil produced by them and its storage, and local transportation, refining and marketing." To achieve this it would "engage in producing, storing, buying, selling, refining, piping, and transportation of petroleum and its products, and the holding of such property as may be necessary for these purposes . . . or any business incidental or auxiliary thereto." The capital of the Producers Oil Company,

Ltd., was oversubscribed to a total of $600,000 before May 1, 1891.[32] The company was formally organized on June 4 representing 1,100 producers.[33]

Sherman was the Trustee who received subscription payments and organized the company. He collected local subscriptions and had printed the book of transfers. He led in the negotiations of the committee that arranged to supply oil to the Oil Creek independent refiners by means of a new pipeline to be constructed to Titusville from the McDonald field in west Pennsylvania. This provided a market for from 8,000 to 10,000 barrels of crude oil per day for members of the Association there. After Sherman insisted, the Oil Creek and Oil City refiners contributed $80,000 to finance the pipeline's construction. As he wrote to the treasurer of the Producers Oil Company, Ltd.: "I have strongly asserted that the Line would be built and that there was but one thing for these refiners to do & that was to get into the omnibus & ride with the rest." [34]

Emery, for some reason, was not included in the board of managers of the new company. An Auxiliary Board was set up, however, to include representatives from producing localities having no representation on the former. As the Trustee, Sherman wrote Emery that he was very anxious that he should be on the Auxiliary Board, "and in your own interest as well as that of the movement accept any position where you can be of use notwithstanding it may not be all that it should be. . . . Do not refuse to take this place, nor to do all that you can to aid the movement. If properly & wisely guided it will be a success." [35]

Sherman advised A. D. Wood, the Treasurer, that the company

[32] *Ibid.,* "Exhibit A," No. 111, containing the statement of policy and the agreement which was presented in typed form as evidence in J. J. Carter *v.* Producers and Refiners Oil Company, Ltd.

[33] Tarbell, *op. cit.,* II, 163–165; Hidy, *op. cit.,* p. 184; Titusville *Sunday World,* November 2, 1895; Williamson and Daum, *op. cit.,* p. 569.

[34] Sherman Papers, Sherman to ?, August 21, 1891, Sherman to J. W. Lee, September 2, 1891, Sherman to Clarence Walker, December 12, 1891, Sherman to A. D. Wood, November 16, December 27, 1891, January 23, 1892. Cf. Johnson, *op. cit.,* p. 173.

[35] Sherman Papers, June 17, 1891.

"should not buy any oil. It is unsafe, speculative and will destroy confidence." Besides, he had it "upon good authority—that the S. O. C. are pickling a rod for that kind of enterprise & for Mellon," who was constructing his Crescent Pipeline to Marcus Hook where he was erecting a refinery. As for a reported "weak tendency on the part of some to 'compromise' with the Standard, there is nothing to compromise. Those people will not regard anything except force and power. These are the weapons they use upon others. . . . The first step towards independence or 'compromise'—whatever the devil that may mean—is to gather strength and power." [36] Such was the lesson that Sherman had learned as counsel of the independents and from his service with the Standard Oil.

As a manager of the National Oil Company he secured from his associates an offer to sell their independent pipeline in the Allegheny County field to the Producers Oil Company, Ltd. When absent from Titusville he delegated to W. W. Tarbell, brother of Ida M. Tarbell, responsibility for securing liberal subscriptions from the Oil Creek refiners to help finance the new pipeline to serve them. All this activity resulted in the construction of a pipeline to Titusville from Coraopolis on the Ohio River, to which the Producers Oil Company, Ltd.'s pipeline extended from the new Allegheny and Washington county oil field.[37]

As early as January 1892 Emery broached to Sherman the possibility of organizing a corporation to construct an independent trunk pipeline from Titusville to New York harbor. At a formal planning meeting Emery promised Sherman that he would be "in" on the project. The latter then threw himself into the work of expediting it with enthusiasm. He consulted Judge Chauncey H. Beasley of New Jersey with a view to securing a law at Trenton granting such a corporation the power of eminent domain by means of a general statute that would also provide for New Jersey regulation and taxation of all oil pipelines, including those controlled by foreign corporations. The producers' "enterprise, and its collateral branches

[36] *Ibid.*, November 16, 1891.

[37] *Ibid.*, Sherman to Wood, December 27, 1891, Sherman to Tarbell, December 28, 1891.

are progressing well," he informed Beasley. "It is as large and important as any in this country, and if brought to a successful termination will be of magnitude sufficient to satisfy professional ambition, as well as those considerations of compensation which are proper to be entertained. I presume your legislature is in session and that no time will be lost." [38]

To Senator Emery, however, Sherman observed: "There are many other seaports to which access is easy and cheap." He could not understand why large capital "should be expended in endeavoring to reach this particular spot on the long shore of the Atlantic Ocean. Baltimore or Philadelphia can be easily reached and Baltimore is preferable, and terminal facilities are plenty and comparatively cheap" there. Some years before, that had been the goal of the producers' Seaboard Pipe Line Company before the Tide Water absorbed it. "As for going to New York harbor, the factor of bucking at the Trust on its own ground should not be permitted to weigh. The objective point is to get into the business and do our share of it. If we can do this we are solid and the hole in the Trust will be just as large." The supply of crude oil and the union of interest between the independent "refiners and producers is the key to the situation," he added, with reference to the alliance of the Oil Creek refiners with the Producers Oil Company, Ltd.[39] Sherman did not intend to be held up for a large sum at Trenton. He wrote Beasley that if the cost of the right of eminent domain proved to be exorbitant the plan to construct the pipeline through New Jersey would be abandoned. However, Sherman and his backers would go to Trenton if necessary to meet with "any persons whom you would recommend us to see." This they did.[40]

Sherman informed Treasurer A. D. Wood of the Producers Oil Company, Ltd., that he would "cheerfully assist you at any time when I can be of use, bearing in mind of course that I am working

[38] Sherman Papers, Sherman to Beasley, January 22, 1892, Sherman to Emery, Jr., January 22, December 23, 1892.

[39] *Ibid.,* Sherman to Emery, Jr., January 22, 1892.

[40] *Ibid.,* Sherman to Beasley, February 13, 1892, three telegrams, Sherman to Beasley, March 1, 1892.

hard for a living and must devote *some* time to that." The "T & O C line," he asserted, "is *the* most important enterprise in connection with the P.O. & P.P.A. *now.* It is important not least in the moral effect it will have." The key to success lay in getting in the capital subscriptions to the Producers Oil Company, Ltd. With arrearages in, the March 80% installment paid, and $74,000 from the refiners, he wrote: "There is absolutely *no* risk in going on. . . . In the meantime the project should be kept alive. People do get lukewarm by inaction & human nature must be dealt with on the plane of its imperfections." Sherman hoped that Wood would "soon get the refiners & *you* and *me* together to perfect a plan of organization & form of contract. Now is the accepted time." They could meet at Wood's "place." The Oil Creek refiners financed forty per cent of the Titusville and Oil City pipeline that connected with the line from Coraopolis.[41]

Sherman had his way. In early March he brought the producers and refiners together in the new Producers and Refiners Oil Company, Ltd., with a capital of $250,000. Of that the Producers Oil Company, Ltd., subscribed $170,000 and the Oil Creek refiners the balance. Sherman had insisted successfully that the former have "both a majority in *interest and number*" in it with a sufficient number of Trustees of the Producers Oil Company, Ltd., on the governing board to overbalance "the *number* of subscribing refiners." [42] The company was formally organized on May 13. J. W. Lee, George H. Torrey, and A. D. Wood were the managers of both companies, which operated as a unit.

Before this, Emery and Sherman had brought the seaboard pipeline project to the fore. This included plans for parallel crude-oil and finished-products pipelines, the latter a distinct novelty. Sherman drafted a contract for signature by the Oil Region refiners obliging them for five years to deliver to the Emery associates' finished-products pipeline all of their refined export oil for transportation to New York harbor at 40¢ a barrel, provided that that

[41] *Ibid.,* Sherman to Wood, February 25, 1892.

[42] *Ibid.,* Sherman to Wood, March 6, 1892; Williamson and Daum, *op. cit.,* p. 570.

pipeline was operating by November 1. Thus they broached formally to the refiners Emery's famous innovation, which would enable the Oil Region independent refiners to market their export oil in Europe in competition with the Standard Oil.[43]

Simultaneously Sherman negotiated authority for the Producers Oil Company, Ltd., to charge an emergency 10¢ pipage should the Standard Oil manipulate certificate and Pennsylvania crude-oil prices adversely. Sherman anticipated realistically the day when the Standard would depress the price of crude oil so as to impoverish the producers while placing a "premium" upon refined when they began to withdraw their oil from control of the Trust.[44] He knew also that the Standard Oil on another occasion could reverse this manipulation.

By this time the complicated project was well advanced. Sherman continued to insist that to free themselves from the Standard Oil yoke the producers had to ally themselves with the independent refiners. Joint investment held the alliance together. As he described the strategy, Sherman declared that it would result in largely increased refining capacity, a larger demand for crude oil, a competitive market, "and consequent better prices." With the addition of a Titusville-to-Oil City pipeline, "the whole scheme can move along harmoniously," taking the crude product under the control of the producers, making profits by transporting it via the pipelines, and marketing the crude product to the independent refiners, first in the Oil Region and the balance ultimately via the seaboard crude-oil pipeline on the coast. To Wood he wrote: "'Don't give up the Ship,'"[45] as the Producers and Refiners Oil Company, Ltd., completed its pipeline from Coraopolis to Titusville and laid lines into

[43] Sherman Papers, undated copy, P. Letterbook IV, 286; Johnson, *op. cit.*, p. 173.

[44] Sherman Papers, Sherman to Clarence Walker, March 16, 1892, Sherman to Wood, March 16, 1892, Sherman to Shannon, March 27, 1892. The refiners who would pay the emergency pipage would retrieve the cost from dividends paid by the two allied companies.

[45] *Ibid.*, Sherman to C. Walker, March 16, 1892, Sherman to Wood, March 16, 1892.

the Butler County oil fields in competition there with the National Transit Company.[46]

While this was being achieved the price of Pennsylvania crude oil sank to 57¢ a barrel. The full force of competition from the Lima and West Virginia fields was now felt in the market which was unprotected by a cooperative producers' shut-down or an arrangement with the Standard Oil Trust. Sherman advised, to counteract the discouragement of members of the Petroleum Producers' Protective Association, that if a "premium" were paid on oil purchased in the fields where the Producers and Refiners Oil Company's pipeline penetrated, the pipeline should "stand the whole of it," but that if certificate oil fell below 50¢ the refiners should pay the emergency pipage so that producers would receive a premium of not exceeding 10¢.[47]

Sherman was exhibiting his old fighting spirit. In late March he recommended that the Executive Board of the Association prod the United States District Attorney of the Southern District of New York to institute antitrust proceedings against the Standard Oil Trust under the Sherman Act. He urged that the Pennsylvania Attorney General be reminded that the Standard Oil's recent absorption of the Atlantic and Western pipelines in Butler and Washington counties violated "the Act of June 15, 1883, Public Laws of Pennsylvania, page 93 &c." He urged that the Association's members patronize better their journalistic friends, the *Sunday World* of Titusville and the *American Citizen,* which he had since sold. Standard Oil Trust representatives were approaching both papers with regard to candidates for the legislature. "*An empty belly makes a weak patriot,*" he informed Wood.[48]

More significantly, in response to an editorial request, he replied to S. C. T. Dodd's "Ten Years of the Standard Oil Trust," published in *Forum* for May, with a savage article, "The Gospel of Greed: The Standard Oil Trust," in the same journal. Afraid of antagoniz-

[46] *History of Butler County* (n.p., 1895), p. 292.
[47] Sherman Papers, Sherman to S. Y. Ramage, March 16, 1892.
[48] *Ibid.,* Sherman to Wood, March 28, 1892.

ing so formidable a business power, the editor cut out the introductory part in which Sherman illustrated and demonstrated "the harmony of the trust methods" of the Rockefeller organization "with certain dangerous tendencies," i.e., the development of "Robber Baron" methods in transportation and the market.[49] Sherman sent the deleted portion subsequently to Henry Demarest Lloyd.[50] The published portion demonstrated the falsity of Dodd's claim that the Standard Oil Trust had lowered refined-oil-products prices significantly.

As for the producers' discouragement, he warned Wood of the folly of talking of building new independent refineries when the refineries allied with them possessed capital of $1,000,000 and the capacity to expand and were being tempted by the Trust, with the bait of lower pipage rates, to break away. He added, vigorously:

> Independence in the business, which means an absolute divorce from Standard support, is what we set out to reach. . . . If men will not give up a penny for the sake of a pound and set a little money in hand over the profits of a boundless future, if they will not broaden themselves out and suppress their pagan natures, then they deserve to be slaves.[51]

Then, on May 2, he sent Emery a preliminary subscription paper for raising capital for the dual seaboard pipeline which was to transport refined export oil as well as crude oil. He advised the Emery associates also that while this paper was circulated the preliminary survey should be made and the right of way secured. As soon as sufficient capital had been subscribed to ensure success and "a reasonable hold on the right of way" obtained, the application for a charter should be made. For the right of way the difficult locations should be first "made sure of." He continued: "If you approve . . . I will get up the papers necessary, & have them all ready."[52] To Sherman the crude-oil pipeline from Coraopolis to

[49] *Forum,* XIII (New York, July 1892), 613–614; Sherman Papers, Sherman to Walter Hines Page, May 5, 23, 1892, MS "The Gospel of Greed: The Standard Oil Trust."

[50] Sherman Papers, Sherman to Lloyd, November 10, 1892, June 5, 1893.

[51] *Ibid.,* Sherman to Wood, March 31, 1892.

[52] *Ibid.,* Sherman to Emery, Jr., May 2, 1892.

Titusville and Emery's seaboard project constituted "a magnificent solution of the problem of independence." "Let both be pushed," he wrote to Lee.[53]

Thereafter Sherman concentrated his efforts upon the seaboard pipeline. After naming the persons who should circulate the contract among the "interior refiners," he resigned from the Auxiliary Board of the Producers Oil Company, Ltd. Of Wood, its Treasurer, he asked: "Is there any way in which a 'Trustee' can resign?" [54] He regularly advised Emery. He helped him with the right-of-way problems. On June 29 he wrote him: "From now to about July 20th it will give me pleasure to dance whenever you fiddle. I will have a paper for incorporation prepared AT ONCE." Sherman drafted the incorporation papers for the United States Pipe Line Company, "the Producers' Line," as it would be known popularly in Butler County. Before leaving for a vacation on Block Island Sherman informed Emery that there was no opposition to his presidency of the new corporation and none to locating its principal office at Bradford if he requested this. As for his own reward, Sherman added: "Take care of my interests. As you know I have worked faithfully in the line for five years, & without reward. I can't afford to work for O. I do not expect any large sums of money, but there is no reason why I should not be employed & paid in Stock—at least in part." [55]

After the United States Pipe Line Company was chartered in September and Emery was elected its President, he employed Sherman on specific, immediate legal problems. The latter continued to advise unofficially on larger problems. When Emery attempted to run his pipeline into upstate New York to connect with the New York, Ontario and Western Railroad after passing under the Erie's right of way with permission, Sherman advised him of the importance of perfecting the right of way "as fast as possible," for which

[53] *Ibid.,* Sherman to Lee, May 3, 1892.

[54] *Ibid.,* Sherman to Wood, June 2, 1892, Sherman to Emery, Jr., May 19, 1892.

[55] *Ibid.,* Sherman to Emery, Jr., June 21, 29, and especially June 16, 1892.

"steady, patient & careful work" was necessary. As counsel responsible for the perfection of title he asked that the company's chief engineer be put to work to secure the detailed data needed for a complete survey, owner identification, and mapping of the route.[56]

Immediately before Christmas Sherman asked Emery for a definite arrangement so that as legal adviser he could have "opportunity to look ahead" instead of answering questions "off hand" that were of great importance to the company but "without knowing their full bearing upon its business." This angered Emery. He replied instantly that he would not use Sherman further because he was unable to give the company's legal business "instant attention." Sherman reminded him immediately of his promise of a year earlier when he had first consulted him about the seaboard project, that he was to be " 'in this.' " Since then, Sherman added, none had "spent more time" or given "more of his energies to the furtherance and organization of plans which led up to this than did I, and it was done unselfishly and without remuneration." He continued:

> It was with the view of putting matters in shape so that I could attend to the business of your company to the exclusion of all other affairs that I wrote to *you* the personal letter of the 20th. . . . Do you not think that you should at least give me an opportunity to act for your company, on some business basis by which I could do it with justice to it and to myself? You say in your letter that you desire this. If so I am ready to undertake to give satisfaction and the responsibility is with you.[57]

A regular arrangement was made. Thereafter Sherman was chief counsel of the United States Pipe Line Company, as he had been unofficially of the Producers Oil Company, Ltd., and the Producers and Refiners Oil Company, Ltd. The $2,000 a year that he received in compensation from the United States Pipe Line Company was grossly inadequate.[58] As Emery's chief law officer he fought the legal battles incidental to the abortive attempt to lay the dual pipeline from Bradford to the New York, Ontario and Western

[56] *Ibid.*, Sherman to Emery, Jr., November 5, 1892.
[57] *Ibid.*, Sherman to Emery, Jr., December 20, 23, 1892.
[58] *Ibid.*, MS Diary 1896.

Railroad. While that attempt was blocked by Standard Oil-induced Erie railroad opposition, and by Standard Oil land purchases, Sherman astutely secured an alternative right of way to Wilkes-Barre to which the two pipelines were laid from Titusville.[59]

At Wilkes-Barre, beginning in June 1893, the United States Pipe Line Company delivered its oil to the Central Railroad of New Jersey. To the delight of the Emery associates the refined oil emerged from its pipeline with quality and color unimpaired. Sherman then labored to secure a right of way through New Jersey to the Columbia Oil Company's terminal at Bayonne.[60] In all this he and his corporate client had the backing of Joseph Pulitzer's *New York World,* which featured the battle of the independent oil men against the Standard Oil. They had, too, the friendly support of the Titusville *Sunday World.*[61] The latter quoted the *Paint, Oil and Drug Review*'s caustic reproach of the Standard Oil's intolerance of rivalry in the oil trade and the ruthless ways in which "it exercises its power to crush competition."[62] The Titusville *Sunday World* also detailed the varied methods employed by Standard Oil agents, namely, enticing local owners to initiate suits against the United States Pipe Line Company and requiring railroads to prevent forcibly the crossing of their rights of way. It reported with pride the independent pipeline's court victories and noted the beneficial effects of its completion to Wilkes-Barre upon the prosperity of the Oil Region cities where the independent refineries served by it were located.[63] A year after that company began pumping oil the *Sunday World* listed twenty-three independent refineries, eight of them in Titusville, seven at Oil City, six in Warren, and two at Bradford

[59] *Ibid.,* Sherman to Emery, Jr., March 20, 21, 1893, affidavit of April 13, 1893, signed by Lewis Emery, Jr., *et al.,* G. J. Clark to Sherman, June 10, 1893; Johnson, *op. cit.,* pp. 174–175; Tarbell, *op. cit.,* I, 168–170.

[60] Sherman Papers, Mercer & Mercer to United States Pipe Line Company, September 12, 1893; Titusville *Sunday World,* March 17, 1894, Williamson and Daum, *op. cit.,* pp. 571–573.

[61] Titusville *Sunday World,* May 26, 1894, quoting *New York World.*

[62] *Ibid.,* March 3, 1894. Cf. Sherman Papers, A. D. Wood to C. Walker, January 7, 1894.

[63] Titusville *Sunday World,* March 3, 17, 1894.

(including the Emery Manufacturing Company), whose aggregate of $5,000,000 capital was giving competition to the Standard Oil, patronage to the independent pipelines, and providing a market for independent producers who again enjoyed a competitive price for their oil.[64] That contrasted sharply, as Sherman knew, with the situation of April 1893 when those refineries had operated part time and lost from five to fifty cents on every barrel of refined oil. Then all the independent interests had fought to live. Now they burgeoned—so great was the success of their attempt to operate businesses in the petroleum industry outside of the Standard Oil Trust.[65]

At Warren on June 4, 1894, J. W. Lee reported the favorable effects of the first year's operation of the United States Pipe Line Company. The price of Pennsylvania crude oil had risen from 58¢ to 87.5¢. The piping of refined oil to Wilkes-Barre and its transportation to the seaboard was overwhelmingly successful. The Oil Region's independent refineries had been sustained despite all the Standard Oil could do to crush them, refining more than a million barrels of crude oil during the year. The unpaid producer-subscribed stock in the Producers Oil Company, Ltd., had been reduced to $20,000. Everywhere the independent oil men were enthusiastic. Such was the annual report of the Producers Oil Company, Ltd., whose officers received $525 in pay each annually, itself evidence of its cooperative character and the integrity of its leaders.[66]

While contributing to the success of the United States Pipe Line Company, Sherman was attacked repeatedly and libelously by the *Oil City Derrick,* whose editor paid similar respects to the company and the independent refiners in a long series of vituperative diatribes. The most amusing canard about Sherman that the *Derrick* circulated was attribution to him of the editorship of a succession of Oil Region newspapers that dared to praise the successful independent oil companies for the contributions they were making to regional prosperity. Sherman advised Emery that the *Derrick's* attacks upon the United States Pipe Line Company and its executives

[64] *Ibid.,* June 23, 1894.
[65] Cf. Sherman Papers, Sherman to Lloyd, April 3, May 8, 1893.
[66] Titusville *Sunday World,* June 9, 1894.

were libelous and damaging. He advised that the *Derrick* be "taught a lesson." He accommodated Patrick C. Boyle, the Standard Oil-subsidized ex-oil scout who edited the *Derrick,* accordingly with a succession of libel suits. The satisfaction with which the Titusville *Sunday World* and its independent oil clientele greeted Boyle's conviction for libeling Emery was enormous. Boyle had to publish a retraction in the *Derrick* and give bond to keep the peace. For his continued libelous attacks on Emery he was rewarded with other convictions and enforced payment of damages to him.[67]

[67] Sherman Papers, Sherman to Emery, Jr., July 10, 1893; Titusville *Sunday World,* April 14, 28, December 8, 1894; Tarbell, *op. cit.,* II, 171–172; Hidy, *op. cit.,* p. 659.

# CHAPTER XIII

# *Climax and Exit*

THE years from 1894 to 1897 brought Roger Sherman's professional career to a second climax. Those from 1878 to 1880 had given him eminence as the Commonwealth's chief associate counsel and the leading counsel of the independent oil men in the Commonwealth Suits and corollary litigation. Now he rose to the peak of his private practice as a corporation lawyer. Technically his position was simply that of chief counsel of the United States Pipe Line Company. Emery's reliance upon him, his close professional ties with the independent refiners of Oil Creek from Titusville to Oil City, and his personal involvement in oil production in western Pennsylvania gave him great influence with the managers of the Producers Oil Company, Ltd., and the Producers and Refiners Oil Company, Ltd. So did his record for astute, farsighted counsel of them during their lengthening fight with the Standard Oil as it shifted from the Trust to the holding company as the structural basis of its monopoly.

These three companies, their 1,100 producer backers, and the independent refiners allied with them, were engaged in a bitter business war with the Standard Oil for the right to compete. Beginning with the blocking of Emery's attempt to lay the United States Pipe Line into New York State to the New York, Ontario and Western Railroad, the Standard Oil fought the Emery associates by every method that it could devise, including the manipulation of the price differential between crude and refined oil while Emery was laying his pipes to Wilkes-Barre. The Standard Oil men also bought into the three allied companies of the Emery group so as to block attempts to combine them and to hamper their cooperation. Buying out individual independent companies, isolating the Emery associates from the Mellon interests, buying in, and "squeezing" in the markets widened the fighting front from New Jersey to the entire realm of independent oil activity in Pennsylvania and on the seaboard. During the long depression of the mid-nineties the crushing of the Pullman strike had put giant business in the ascendancy. The Standard Oil's counterattacks were business war in the classic manner of the so-called "Robber Barons" who were determined to exterminate competition and organize entire provinces of business into separate monopolies.

Some of the bitterest fighting, involving open resort to violence, occurred in New Jersey. It was through that state that Roger Sherman secured the right of way for the United States Pipe Line Company that would carry its pipelines from Wilkes-Barre to the oil docks of the Columbia Oil Company at Bayonne on New York harbor. The request of the Standard Oil that its railroad friends protect its interests induced the Lackawanna, one of whose directors was a Standard Oil man, and the Pennsylvania to oppose laying the dual pipeline under their rights of way.[1] Ironically, not only the Tide Water Pipe Line but also the Standard Oil's crude-oil pipelines extended under these same railroads as they crossed New Jersey to New York harbor. The railroad opposition was heightened by the fact that Emery was shipping his oil from Wilkes-Barre to Bayonne via the Central Railroad of New Jersey. To remove this obstacle Sherman persuaded the Emery associates to press their request for a free-pipeline law like Pennsylvania's at legislative hearings at Trenton. Simultaneously, Sherman acquired a farm with a right of easement via a culvert under the Lackawanna tracks for the pipeline right of way.

For all the astuteness of Sherman's strategy he and Emery were defeated, not without exhibiting sterling fighting ability on foot and in the courts.

In the legislative hearing all seemed favorable for the free-pipeline bill as Emery, Lee, and others of the associates stated its advantages and their interest in its enactment. A senator agreed to

[1] Tarbell, *op. cit.*, II, 183–185.

present and sponsor it. No opposition appeared until the closing days of the legislative session, when suddenly an influential hostile lobby mobilized, and the senatorial sponsor quietly left for the far west where he remained until after adjournment.[2] Not for nothing was New Jersey the "Mother of the Trusts."

Meanwhile, holding the farm in fee simple with its right of easement under the Lackawanna's tracks, Emery mobilized his construction crew and laid his dual pipeline in the culvert, protected by armed guards. The Lackawanna Railroad immediately counterattacked on October 28, 1895, with three wrecking trains and 200 to 300 men. Emery put himself at the head of his men and successfully defended his newly laid pipeline with the aid of petroleum torches along it. A pitched battle drove the railroad men off; Emery maintained his armed guards there for eight months. Then Sherman applied for an injunction from Vice-Chancellor V. C. Bird at Trenton to restrain that carrier from interfering with the pipeline. Afterwards F. F. Murray's dispatch to the Titusville *Sunday World* observed that there were no cannon opposing Emery as there had been when he had attempted to lay pipes under the Erie's tracks at Hancock, New York, two years earlier, and that although Standard Oil badges were lacking, the railroad forces again were impelled by its influence.[3]

Vice-Chancellor Bird decided, however, that the United States Pipe Line Company's action in laying its pipes in the culvert without the Lackawanna's permission was untenable. He refused to issue the injunction. Sherman was positive that Emery's position was correct and appealed the case. Meanwhile, the Pennsylvania Railroad had secured a temporary injunction prohibiting Emery from laying his pipeline under its roadbed. Sherman appealed from that also. In default of the free-pipeline law he had to fall back upon the precedent provided by other, earlier pipelines and the fee-simple right of easement under the Lackawanna.[4]

[2] *Ibid.*, p. 186.

[3] Sherman Papers, Lloyd to Sherman, December 30, 1895, Emery to Lloyd, January 4, 1896; Titusville *Sunday World*, November 2, 1895.

[4] Sherman Papers, Emery to Lloyd, January 4, 1896, Opinion of V. C. Bird, MS 1896 Diary, January 3, 1896; Lloyd Papers (Winnetka), Sher-

Meanwhile, the powerful Mellon interests of Pittsburgh had completed the Crescent pipeline to refineries situated at Marcus Hook on Delaware Bay. This operated for a short while before the high price of crude oil and the low price of refined oil in early 1894 and the undercutting of their European market obliged the Mellons to sell out to Standard Oil interests. While this isolated the enterprises of the Emery associates soon after the United States pipeline was completed to Wilkes-Barre, the Standard Oil was defeated temporarily in its attempt to combine the Crescent line with the National Transit Company by Governor Pattison's veto of the repeal of the act of 1883 which forbade the consolidation of competing pipelines.[5]

Early in 1894 the Standard Oil attempted to buy out the independent refiners seriatim, together with their shares in the Producers Oil Company, Ltd., the Producers and Refiners Oil Company, Ltd., and the United States Pipe Line Company while the depressed price of kerosene kept the refiners under pressure. Late that year three leading Titusville refining companies with which Sherman had worked closely were obliged to sell out to the Standard Oil, thereby diverting their business from the refined-oil pipeline of the United States Pipe Line Company.[6] Almost simultaneously, the National Transit Company refused to connect with the wells of independent producers who were beyond the reach of the independent pipelines. This, like the attempts to block the completion of the United States pipeline, led to expensive litigation. In this instance Sherman was able to secure a court order ordering the National Transit Company to connect with those wells.[7]

Henry Demarest Lloyd plied Sherman with queries regarding the independents' litigation and problems during those years. Sherman's

---

man to Lloyd, January 4, 1896; United States Industrial Commission, *Preliminary Report on Industrial Combinations,* I, 103–104, testimony of Lewis Emery, Jr.; Johnson, *op. cit.,* pp. 180–182.

[5] Johnson, *op. cit.,* pp. 175–176; Williamson and Daum, *op. cit.,* pp. 582–585.

[6] Lloyd Papers (Winnetka), Sherman to Lloyd, February 17, 1894; Johnson, *op. cit.,* p. 178.

[7] Sherman Papers, A. D. Wood to C. Walker, January 7, 1894; Lloyd Papers (Winnetka), Sherman to Lloyd, February 17, 1894.

replies, some of them of considerable length, provide insight into the situation and understanding of his role as chief counsel of the United States Pipe Line Company. As a lawyer he was fully alert to Standard Oil and railroad encroachments upon the independents' legal rights. He agreed with Lloyd that "attempted lawful competition with lawless competition will result in the destruction of the weaker." [8]

Early in 1894 Sherman had advised Emery that a consolidation of the independent pipeline companies was essential for survival. The United States Pipe Line Company then proposed to purchase the pipeline of the Producers and Refiners Oil Company, Ltd., which supplied crude oil to Titusville and Oil Creek refiners. The Standard Oil countered by buying stock in the Producers Oil Company, Ltd., in an attempt to prevent this. Further, after completion of the United States pipeline to Wilkes-Barre, the Standard Oil depressed the price of export refined oil for German markets to 2¢ F.O.B. New York City so as to undercut the Emery associates' new overseas market. To French refiners, whose supply of crude oil the Standard Oil monopolized, it sold crude oil at 3.6¢ per gallon F.O.B. This Senator Emery discovered when he was abroad in September. Sherman was critical of the independent refiners' neglect of the domestic market, which made them vulnerable to the pressure that such overseas destructive price cutting produced. However, the New England Oil Company was similarly undercut in its regional market by prices of 4.75¢ and then 1¢ per gallon on delivered illuminating oil! The Standard Oil thus countered with cutthroat competition wherever the independents offered to compete.[9] It was the heavy loss on refined export oil that broke the backs of some inde-

[8] Sherman Papers, Lloyd to Sherman, February 5, 1895, Sherman to Lloyd, February 25, 1895; Lloyd Papers (State Historical Society of Wisconsin), Sherman to Lloyd, February 25, 1895.

[9] Lloyd Papers (State Historical Society of Wisconsin), Sherman to Lloyd, February 25, 1895, Emery, Jr., to H. A. Drury, September 28, 1894; Titusville *Sunday World,* November 24, December 8, 1894; Ida M. Tarbell Papers (Allegheny College Library), W. W. Tarbell to I. M. Tarbell, May 6, 1894; Johnson, *op. cit.,* p. 178; Tarbell, *op. cit.,* II, 173–174, 212.

pendent refiners who then sold out to the Standard Oil late in 1894.

In mid-March 1894 J. J. Carter, the wealthy ex-tailor of Titusville who was now a successful producer, sued for two injunctions against the sale of the Producers and Refiners Oil Company pipeline to the United States Pipe Line Company. Sherman represented the Producers and Refiners Oil Company, Ltd., in the ensuing litigation, a case which attracted wide interest. The Allegheny County Court of Common Pleas upheld Carter's objections and granted a preliminary injunction on the ground that the managers of the Producers and Refiners Oil Company, Ltd., lacked the authority to sell the pipeline in exchange for stock in the United States Pipe Line Company and that the relevant bylaw had not been observed in polling the members of that limited partnership.[10] When the same issue was argued at Meadville in the Crawford County Court of Common Pleas, two Pittsburgh attorneys headed Carter's five counsel in opposition to Sherman and two associates. It was evident that Carter represented others than himself in the action.[11] As President of the Carter Oil Company he held 300 shares of the Producers Oil Company, Ltd., which controlled the Producers and Refiners Oil Company, Ltd. His petitions for injunctions obliged these independent companies to divulge their interlocking relationship. The case was argued at the time when Coxey's Army was marching upon Alliance, Ohio, causing widespread apprehension. Although it was elicited from Carter that the Standard Oil owned 60% of the stock of his oil company, despite Sherman's utmost efforts a preliminary injunction was issued against the sale by Judge John Henderson, who cited the similar decision in Carter *v.* Producers Oil Company, Ltd. Sherman appealed to the Supreme Court. As an ally of the Standard Oil, Carter had succeeded in his attempt to prevent the merger of the three independent pipeline companies which Sherman urged and Emery intended to accomplish.[12]

[10] *Titusville Morning Herald,* March 16, 26, 1894.

[11] *Ibid.,* March 16, 28, 1894.

[12] *Ibid.,* March 28, 29, 30, May 4, 1894; Sherman Papers, MS affidavit of A. D. Wood in John J. Carter *v.* The Producers Oil Company, Ltd., J. W. Lee *et al.,* March 16, 1894; Tarbell Papers, W. W. Tarbell to I. M.

The independent producers were infuriated when Carter appeared at the Warren meeting of the Producers Oil Company, Ltd., on April 11 with 13,013 shares that had been loaned to him by the Standard Oil. Despite the *Titusville Morning Herald*'s defense of Carter's objections against consolidation 413 members voted for this against Carter's minority of 27 in accord with the voting regulations. The Petroleum Producers' Protective Association promptly ousted Carter from its membership.[13] When he opposed dissolution of the Pittsburgh injunction against the sale of the Producers and Refiners Oil Company pipeline Sherman appealed that to the Supreme Court also.

Subsequently, despite Sherman's argument that the majority of the members of the Producers and Refiners Oil Company, Ltd., possessed the undoubted right to sell their company's pipeline to the United States Pipe Line Company, and that all of the stockholders present at a special meeting had voted to do so, the Pennsylvania Supreme Court upheld the injunction. The court ignored his plea that Carter was not a stockholder in nor a member of the former company. It reserved the larger issues involving the character of limited partnership associations, which Sherman had raised, for a future decision. Although the Producers Oil Company, Ltd., in which Carter held stock, did own 17/25's of the stock of the Producers and Refiners Oil Company, Ltd., Sherman's impatience with the Supreme Court's decision is very understandable.[14]

By late October 1894 it became obvious therefore that, as the Supreme Court of Pennsylvania interpreted the law of limited partnership associations, the Emery associates were unable to merge their three independent pipeline companies, or their pipelines, for-

---

Tarbell, May 6, 1894; Tarbell, *op. cit.*, II, 178; Hidy, *op. cit.*, pp. 272–273; Johnson, *op. cit.*, p. 178.

[13] Lloyd Papers (Winnetka), Sherman to Lloyd, June 8, 1894; Tarbell Papers, W. W. Tarbell to I. M. Tarbell, May 6, 1894; *Titusville Morning Herald*, April 13, 1894; Johnson, *op. cit.*, p. 178.

[14] *Pennsylvania Reports*, CLXIV, 463–469; Sherman Papers, *Paper Books, XXVI, John J. Carter, Appellee* v. *Producers and Refiners Oil Company, Limited, et al., Supreme Court of Pennsylvania, Eastern District, No. 82, July term, 1894;* Tarbell Papers, W. W. Tarbell to I. M. Tarbell,

mally over the objections of the pro-Standard Oil minority stockholders. Carter then purchased the Standard Oil-owned shares in the Producers Oil Company, Ltd., and attemped as majority owner to control its management. This was frustrated when the managers refused to register the stock transfer. That action the courts upheld. The Producers Oil Company, Ltd., then bought in his shares. Carter, however, was victorious in his attempt to prevent the consolidation that Sherman had sought.[15]

The Standard Oil tactic of buying into the independent companies was attempted with regard to the United States Pipe Line Company, also. In that instance the courts eventually required that its management permit the Standard Oil representatives to vote their shares and be represented upon the Board of Directors.[16]

Consolidation of the independent companies for the better marketing of their crude and refined oil was still imperative. Indignant producers' meetings had purged the managers of the Producers Oil Company, Ltd., of "traitors" in the spring of 1894. The depressed price of kerosene convinced the managers of the Producers and Refiners Oil Company, Ltd., that the Standard Oil intended to "crush out" the independent refiners. A libel suit that Sherman launched against the *Morning Herald* for misrepresenting the authorized capital increase of the United States Pipe Line Company hardly solved the problem. He had foreseen at the outset the possibility of a scissors tactic in the markets when the second independent producers' pipeline company was organized. To weather the crisis the Producers and Refiners Oil Company, Ltd., had to curtail salaries and staff. William W. Tarbell, one of its officers, anticipated a successful consolidation. That did not materialize for some years.[17]

---

May 6, 1894; *Titusville Morning Herald,* May 4, 1894. The members of the Producers Oil Company, Ltd., had also voted for the sale.

[15] Sherman Papers, MS 1896 Diary, March 4, April 16, June 19, 22, 25, 1896; Tarbell Papers, W. W. Tarbell to I. A. Tarbell, June 9, 1896; Tarbell, *op. cit.,* II, 179–181; Hidy, *op. cit.,* p. 646.

[16] Tarbell, *op. cit.,* p. 181.

[17] Sherman Papers, Sherman to A. D. Woods, March 16, 1892; Tarbell Papers, W. W. Tarbell to I. M. Tarbell, May 6, July 18, 1894; *Titusville Morning Herald,* May 8, 1894.

Meanwhile, the Titusville *Sunday World* declared in a feature article, "The Independent Oil Men," that they were marching toward "the overthrow of the most powerful and unscrupulous monopoly in this country." Ex-Senator J. W. Lee, chairman of the managers of the Producers Oil Company, Ltd., added that the public should support the independents because the Standard Oil Trust was contrary to the common law and was parent of "the great brood of trusts which have secured such a grip on this country." A return in the petroleum industry to "the fair and natural laws of competition," the *Sunday World* declared, would make "calling of other trusts to account . . . a comparatively small matter." Thus the cause of the Emery associates was the "test case for the people against the Trusts—and sooner or later it is bound to win." [18] After the Pennsylvania Supreme Court's decision in the Carter case the *Sunday World* declared that the public must face the fact that existing law might be inadequate to cope with "the parent of this great brood of trusts" and must act accordingly or else free, democratic government would fail.[19] This hint that additional antitrust legislation was needed, foreshadowed the view of the reformers of the Progressive Era, which would subsequently produce the Clayton Act. This also paralleled Sherman's advice to the Emery associates that "the great underlying question" involved in the struggle with the Standard Oil must be fought out in the courts.[20]

Sherman was insistent upon effective combination of the independents so as to present "a united front" and "a body strong enough to make any kind of fight necessary." This was especially to enable them to withstand the market pressure resulting from adverse crude- and refined-oil price changes. Emery went abroad with a plan for the international association of the independent oil interests of the United States, Germany, Great Britain, and France, including tanker owners, to be organized under the copartnership

[18] Titusville *Sunday World,* June 9, 16, 1894.

[19] *Ibid.,* December 29, 1894.

[20] Lloyd Papers (State Historical Society of Wisconsin), Sherman to Lloyd, February 25, 1895.

laws of Germany,[21] and at the same time he cultivated overseas markets for the export and crude oil of the Emery associates.

Sherman then presented his plan to A. D. Wood in detail. All the independent companies of the Emery associates, he felt, should be operated as a single enterprise. Had this been done from the beginning their large resources would have enabled them to secure financial assistance. "As it is now, and as it has been throughout, diverse interests were continually asserting themselves selfishly, and interfering with the profitable operations of any one." Furthermore, the exporting policy had broken the refiners' backs while the home trade, which was to be developed by the refined-oil pipeline, "has been totally neglected." The only way to save the independent enterprises, he said, was to "*consolidate.*" Although two years had been lost, nevertheless to do so would produce great financial strength "even now." The problem was how to do it in time, and how to muster the "utmost energy" to accomplish this. Instead of wasting time in meetings, Sherman suggested sending three to five "resolute men" to present the case for consolidation as the only remedy to individuals. The only alternative, he reminded Wood, was the United States of America completely subjugated by the Standard Oil Trust.[22] This statement was provoked by the limited success of the Standard Oil's campaign to purchase the independent refiners and by its attempt to pit the producers against the survivors.

Emery visited all the major European trade centers, called upon the German Imperial Court, and then went to St. Petersburg. Although his projected international association of independents did not materialize, he established friendly relations with the Nobel interests and secured Dr. Philip Poth of Mannheim as the chief marketing agent of the United States Pipe Line Company and the Pennsylvania independent refiners abroad. Aside from the Standard Oil's ruthless price competition, the foreign-market position of this group seemed to be assured.

[21] Sherman Papers, Emery, Jr., to Dr. Philip Poth, August 18, 1894, Sherman to Lloyd, February 25, 1895.

[22] *Ibid.,* Sherman to A. D. Wood, November 21, 1894.

On January 24, 1895, a general meeting of the independent oil producers was convoked suddenly at Butler, Pennsylvania, under the auspices of the managers and officers of the Emery associates' three pipeline companies. Their achievements, their problems, and the Standard Oil's hostile methods were discussed candidly. Especially pointed was the disclosure that the crude-oil equivalent of a barrel of kerosene then cost 34¢ more than the latter. Wood attributed this solely to the Standard Oil's attempt to break down all competition and to obtain an absolute monopoly of the business."

J. W. Lee, the chairman, and Emery urged the now-prospering producers to sustain their program and attempt to restore competition. They told the producers that they were "in the strongest position" that they had been in "for the last sixteen years," since they were receiving $600,000 *more* per year for 80,000 barrels of crude oil per day than they had received when they produced 120,000. Determined to assert their "God-given right to do business" they adopted Emery's resolution pledging them to sustain financially "all the independent oil interests with which they have or may become identified, in producing, transporting and refining." Whereupon they subscribed $50,000 to sustain the hard-pressed independent refiners, and increased this to $200,000 later. This was their reply to the Standard Oil's offer of fifteen days earlier to buy all the independent pipelines and refiners "and pay them what they are worth." [23]

Sherman was much encouraged. He talked with the representatives from all districts represented there and was convinced that the independent refiners would be sustained to the extent of $500,000. On the basis of this, nearly all of the independent refiners would be willing to continue. The Oil City refiners present "expressed themselves satisfied." So he wrote to Fertig,[24] who, however, sold his refinery to the Standard Oil.

After Sherman left for Washington, David Kirk, a most active

[23] *Titusville Morning Herald,* January 25, 1895. Cf. Tarbell, *op. cit.,* II, 175–176.

[24] Tarbell, *op. cit.,* II, 175–176; Sherman Papers, Sherman to John Fertig, January 26, 1895.

and influential leader, secured the general meeting's enthusiastic endorsement for the organization of a new company to be called the Pure Oil Company, which would aid the refiners in their predicament. Some $200,000 was subscribed toward it at once. The new company was intended to be the financing and marketing agency of the three independent pipeline companies and of the independent refiners, with $2,000,000 authorized capital. It would, the *Pittsburgh Post* declared, "complete the chain of independents between the producer and consumer" and remedy existing defects in "the machinery for competing with the Standard." [25]

The Pure Oil Company's organization was completed in November. So bitter were they at the adulterated oil being marketed by the Standard Oil that the dealers' response was very friendly. The new company's outlook was declared to be very bright by Kirk, its President. That the Pure Oil Company was the practical fruit of Roger Sherman's insistence upon the consolidation of the independents' enterprises may be questioned, since its birth increased their number. It capped the pyramid of the independent enterprises in production, refining, pipelines, and marketing.[26]

Henry Demarest Lloyd followed this development closely. Soon after announcement of the plans for the Pure Oil Company he informed Sherman that he could not see how the independents "can hope for success if they confine themselves to competition." He continued: "Unless these Trust men can be brought to justice, every competitor will be at the mercy of their loaded dice. Tweed, all powerful as he seemed, was brought to justice. Why not Rockefeller?" [27]

Sherman replied that he agreed that it was now time to "contest the great underlying question," that of the Standard Oil's illegal monopoly in restraint of trade under the Sherman Act. "In no other case can the facts be made as strong, and in no other instance are facts so well known and so well marshaled. I have repeatedly urged this upon our people orally and in writing." He hoped that before

[25] *Titusville Morning Herald,* February 9, 1895, quoting.

[26] Sherman Papers, Sherman to Lloyd, February 25, 1895.

[27] *Ibid.,* Lloyd to Sherman, February 21, 1895.

the independents "get through, they will give the Standard Trust a legal battle on a large scale." He added: "We could make of the Standard Trust such an object lesson for the world to gaze upon as would produce results, even, if, in the end, we were absolutely defeated." [28]

The Standard Oil's reply to the independents' new plans was to push the price of Pennsylvania crude oil higher. Aided by speculative buying and a temporary decline in oil stocks, this sent the price to $1.27 a barrel on April 10, $2.25 on April 17, and ultimately to $2.60. The independent pipelines had to buy their customers' oil at these abnormally high prices and store it in large quantities. At the same time they were obliged to sell the product of the independent refiners at the abnormally low market price fixed by the Standard Oil. Thus, while the speculative market in crude oil enabled producers who had held on to 93¢ oil and speculators in certificates to make "good-sized wads" as the *Morning Herald* observed, the total effect was to intensify the market squeeze upon the independent refiners.[29]

It was during this great squeeze that the Pennsylvania legislature repealed the anticombination pipeline law of 1883 which action enabled the National Transit Company to complete the acquisition of the Crescent pipeline and finally isolate the Emery associates. The latter regarded this action as "a disastrous blow" to the independent pipelines, opening the way to constant open and covert attacks from "the single monopoly that rules the oil business." [30] After the Crescent line passed into possession of the National Transit Company the price of crude oil sank to $1.25, a price level which producers attributed to the Standard Oil but which was also related to a substantial increase in production.[31] Meanwhile, all but three of the Titusville refiners had had to sell out to the Standard Oil.[32]

[28] *Ibid.,* Sherman to Lloyd, February 25, 1895.

[29] *Ibid.,* Sherman to Lloyd, February 2, 1897; *Titusville Morning Herald,* April 10, 18, 20, 1895.

[30] Sherman Papers, circular dated January 26, 1895.

[31] *Butler Citizen,* August 1, 1895.

[32] Lloyd Papers (State Historical Society of Wisconsin), Sherman to Lloyd, July 16, 1895.

Sherman was bitter in his opposition to the Standard Oil. In July 1895 he wrote Lloyd that the abuse of individuals by the Standard Oil-controlled press was "only hastening the bitter ending" of that Trust's "career of crime." [33] In January 1896, after Vice-Chancellor Bird at Trenton, New Jersey, had rejected his request for an injunction restraining the Lackawanna Railroad from interfering with the United States Pipe Line, he predicted that soon:

> [It would be] a matter for astonishment and ridicule that a judge could be found to decide . . . that a monopoly should be encouraged by forbidding one transportation line to do what another has been permitted to do, that is, lay a pipe line under railroads, and that . . . a grant made by a state for public purposes . . . to facilitate commerce among the people, could afterwards be used by the railroad company to impede and obstruct commerce through the state.[34]

Four days after Vice-Chancellor Bird's adverse decision, Sherman met with Emery, Lee, and others of the Emery associates at the Belvidere House in Buffalo. Their consultations concluded with a decision to incorporate the Pure Oil Company as a New Jersey corporation. "Big stick," Sherman noted in his diary that night.[35] A fortnight later he was in Bradford for a United States Pipe Line Company conference. Six days after that he was in New York City, Newark, and Jersey City, conferring with Emery and arguing cases arising from that pipeline's attempt to make good its entry into the New York harbor area.[36]

Such activity was typical of Sherman's life during those months. He gathered data on John J. Carter's relation to the Standard Oil. On March 4, 1896, Carter informed him that he owned a majority of stock of the Producers Oil Company, Ltd., and that he was "with S. O. Tr., in it for the money there is in it." [37] This was before the Producers Oil Company, Ltd., bought him out. On the next day at Newark, New Jersey, Sherman defended the United States Pipe

[33] *Ibid.*

[34] Lloyd Papers (Winnetka), Sherman to Lloyd, January 4, 1896.

[35] Sherman Papers, MS 1896 Diary, January 8, 1896.

[36] *Ibid.,* January 21, 27, 1896.

[37] *Ibid.,* March 4, 1896.

Line Company before Vice-Chancellor V. C. Emery in Delaware Lackawanna and Western Railroad *v.* United States Pipe Line Company. Then he conferred with Emery and others on United States Pipe Line Company matters until dinner, after which he entrained for Philadelphia. There he gathered data on how to fight Carter's bid for managerial control of the Producers Oil Company, Ltd., calling upon Federal Judge Acheson, "Mr. Hopkins at the People's Bank," Heydrick, and a Judge McMichael of the new Court of Appeals.[38] On this occasion he went to the theatre and saw "Southern in 'Prisoner of Zenda' Broad St Theater," as he recorded in his diary. The next morning he left for Pittsburgh where business kept him only part of a day before he entrained for Titusville.

The Producers Oil Company, Ltd., which Carter and the Standard Oil attempted vainly to control, stopped the downward trend of crude-oil prices by paying a premium of 7.5¢ a barrel in the market. This obliged the Standard Oil to raise its pegged price above $1.20.[39] Sherman was hurt when at a meeting of the directors of the United States Pipe Line Company on April 16 another attorney, D. T. Watson, was appointed to take charge in "Carter's suit," and he was instructed to go to New Jersey and manage the company's litigation there. "Lee & Co. pass me by and I think intentionally sometimes," he wrote in his diary. Lee was now Vice-President of the Pure Oil Company.

Not unrelated to the independents' fight for competitive free enterprise in the petroleum industry was T. W. Phillips' introduction in Congress of a bill to establish an Industrial Commission of twenty-one members representing equally agriculture, business, and labor. In his successful championship of this measure Phillips urged that "the industrial issue" be met by building upon Jeffersonian principles. "The State should guarantee each individual the right to pursue happiness," he said, and organize its "social and industrial system upon a more just and equitable basis." The proposed Commission would "give an impartial hearing to those who complain of discriminating laws and unequal burdens." In this manner the inde-

[38] *Ibid.*, March 5, 6, 1896.

[39] *Butler Citizen,* April 16, 23, 30, May 14, 1896.

pendent producers' movement, with Congressman Phillips as its spokesman, again promised to influence the course of national development. Before the Commission the hard-pressed independent oil men would have an opportunity to present their cause and its problems and the issue of competition versus monopoly, and would thus influence national policy. This they did, two years after Sherman's death, with Phillips presiding as the Chairman.[40]

On May 16 Sherman's diary contains a single reference. "J. W. Lee in Office." This visit was related undoubtedly to the Standard Oil's sudden success in tricking Phillip Poth into selling out to it, by informing him that the American independents were also selling out to Standard Oil. Thus the United States Pipe Line Company suddenly lost its continental distributor, to whom 700,000 barrels of independent refined oil had been shipped annually for three years. Having been guaranteed $200,000 to finance the replacement of Herr Poth, on July 4 Emery sailed for Germany on a secret mission in an attempt to win him back or to open up another channel for marketing independent oil in Europe. Poth died before he arrived. Emery was obliged to organize Pure Oil Company distributing centers in Germany and Holland.[41]

Shortly before Emery sailed, Sherman had figured as a witness only in Carter *v.* Producers Oil Company, Ltd., at Warren. After Watson was named defendant's attorney, Sherman had refused repeatedly to assist in the defense. The case was important, as William W. Tarbell informed his sister Ida, since Carter was attacking the constitutionality of the Pennsylvania limited-partnership statute. That affected "an immense amount of capital organized under the law." It alienated from Carter many of his best friends, who called him "a fool." [42]

As a step toward consolidation Sherman advised Lee that the

[40] *Ibid.,* May 28, June 4, 11, 1896; United States Industrial Commission, *Preliminary Report,* I, *passim.*

[41] Sherman Papers, MS 1896 Diary, May 16, June 4, 30, 1896; Lloyd Papers (Winnetka), Emery, Jr., to Lloyd, June 3, 1896; Tarbell, *op. cit.,* II, 177.

[42] Sherman Papers, MS 1896 Diary, June 15–25, 1896; Tarbell Papers, W. W. Tarbell to I. M. Tarbell, June 9, 1896.

Producers Oil Company, Ltd., should sell its United States Pipe Line Company stock to the Pure Oil Company.[43]

In early July Sherman won the appeal from the lower New Jersey court injunction restraining the United States Pipe Line Company from laying its tracks under the Pennsylvania Railroad. He concluded sanguinely that this "practically enables the Pipe Line Co. to complete its line through to the seaboard, as I believe that the Lackawanna case (also on appeal) will be shortly decided favorably to the Pipe Line." Happy in anticipation of this prospective victory—illusory because both cases were carried on appeal up to the New Jersey Supreme Court of Errors and Appeals—he scheduled a vacation for Block Island and informed Lloyd, "I should like very much to meet you while I am there." [44]

After many delays (and it was after Sherman's death) the United States Pipe Line Company lost both of these appeals, inexplicably were it not for the influence of the Standard Oil Company of New Jersey. After that defeat Emery had to pull up his pipes. He turned, perforce, to Pennsylvania where the free-pipeline law enabled him to secure a right of way to Marcus Hook. There he acquired at cost a terminal on Delaware Bay purchased earlier by an associate, completed his dual pipeline to salt water at that point, and on May 2, 1901, pumped there the first refined oil ever to reach the coast by pipeline.[45]

Sherman and Lloyd conferred together on Block Island, to which Lloyd and his sons had sailed in their sloop, on the European market crisis confronting the Pure Oil Company and the United States Pipe Line Company. The tenor of Lloyd's advice can be gleaned from a letter to Sherman that he had written two days earlier. The trouble with the independent oil men, he wrote then, "as with all the other victims of the trusts, is that they are not radical enough." He continued: "Our industrial usurpers know no such half-heartedness. They are pushing their advantage . . . and using without stint

[43] Sherman Papers, MS 1896 Diary, June 30, 1896.

[44] Lloyd Papers (Winnetka), Sherman to Lloyd, July 7, 1896.

[45] Tarbell, *op. cit.,* II, 187–188.

every political power as well as every industrial one to the utmost." America, he said, should have "every monopoly of necessaries of life promoted to a government function, and we should have had our Rockefellers, Carnegies, Vanderbilts and Goulds either established as government heads of industrial administration, or incarcerated in the penitentiaries for their frauds."[46] Sherman undoubtedly preferred the latter alternative. While on Block Island, he paid $500 into the Pure Oil Company's "European fund" of $200,000 to finance its marketing program abroad. He had already subscribed to Pure Oil common stock.[47]

After his return to Titusville Sherman spent his entire time preparing to defend the United States Pipe Line Company from the attempt of the National Transit Company to vote its stock in it and thus intervene in that independent's affairs. On August 17 he conferred at Bradford with T. W. Phillips, Lee, and Hugh King, President of the Columbia Oil Company, regarding the matter. They authorized him to file an answer asserting that the Standard Oil subsidiary had no right to hold stock in another company or to vote by proxy. They also instructed him to file a cross bill in the United States District Court of New York City for an injunction against the National Transit Company's attempt to do so and also to petition the Pennsylvania Attorney General for cooperation in applying for a writ of *quo warranto* against the National Transit Company from the Supreme Court of Pennsylvania.

From August 24 to August 27 Sherman attended the three-day conference of the executives of the Pure Oil, United States Pipe Line, Producers Oil, and Producers and Refiners Oil companies with certain French and German oil men at Long Branch, Long Island. Seventeen met daily in this conference. Emery, Phillips, Kirk, Lee, Hugh King, and Byles, Sherman's neighbor, were there. Swimming parties were interspersed between the meetings. On the afternoon of the second day Sherman drove in a carriage with Emery to discuss specific refractory problems. "Seur, Walz, Liebel," as Sherman noted

[46] Sherman Papers, Lloyd to Sherman, August 6, 1896.

[47] *Ibid.*, MS 1896 Diary, June 30, July 28, August 8, 1896.

in his diary, represented France and Germany in the sessions. The conference nearly split on the third day. "General appearance *bad,*" Sherman entered in his diary on the issue of consolidation. Lee, Sherman, and E. H. Jennings were appointed to a special "Consolidation Committee." Jennings had come late. Sherman interviewed Emery and Phillips "on Consolidation & Pure Oil Co." on the piazza. Then, before the conference adjourned, Sherman was granted "power" by the conference. That included not only consolidation but also authority to launch criminal antitrust proceedings against the Standard Oil by means of a bill of complaint. Possibly it was this that he discussed with Hugh King in his Columbia Oil Company office that afternoon in the presence of Byles and Emery.[48]

Subsequently, as Emery's guest at Bradford from August 31 to September 1, the business attended to by Sherman in a series of conferences with Emery, King, and "Collins" was, according to his diary:

Wood Stock Trusts
Trust agreemt (2)
Pure Oil Co Trust
Answer N. J. Suit
Corp bill—Counter Suit
French & English Contracts

That early he was urging the transformation of the Pure Oil Company into a trust which could consolidate the three pipeline companies with itself by holding their stock in trust. This would integrate control of oil wells, pipelines, refineries, and marketing facilities and achieve thereby the strength that Sherman insisted was essential for survival in competition with the Standard Oil. Laying the foundation for this at the Bradford conference was undoubtedly his greatest achievement in corporation practice. The Long Branch and Bradford conferences also reopened European markets for the independents' products in Great Britain, France, and Germany. Then, on September 16, in pursuance with a third decision made at Bradford,

[48] *Ibid.,* August 17, 24–27, 1896; Tarbell, *op. cit.,* II, 186.

he filed the defendant's answer in National Transit Company *v.* United States Pipe Line Company.[49]

As if in acknowledgment of Sherman's extraordinary legal acumen Patrick C. Boyle attacked him characteristically in the *Oil City Derrick.* When Lloyd called his attention to this Sherman replied: "I never see the miserable sheet. It and its editor are entirely beneath contempt . . . you do not need to bother yourself with him, nor anything that he says. He keeps his unique position as a slanderer for a Trust by showing extraordinary zeal." Boyle, Sherman added, was "a mere hireling of a band of thieves, he is entirely too low to bother with." Such was his estimate of the Standard Oil executives and their journalistic libeler.[50] Sherman knew how his personal friend N. M. Allen had been ruined by the Standard Oil as a refiner and publisher in the mid-seventies and been persecuted by it ever since because he had exposed the South Improvement Company in late January 1872.[51]

In between repeated train trips to Smithport, Pennsylvania, to attend the preliminary hearings of National Transit Company *v.* United States Pipe Line Company, or to New York City where he took testimony relevant to the case at the office of John McClure, 22 William Street, Sherman went to Bradford to confer with Emery or to see his own physician, a Dr. C. S. Hubbard. On such trips he met such Standard Oil notables as Joseph C. Sibley of the Trust, garbed in a beaver coat and a "prehistoric silk hat," Joseph Seep of the Seep Purchasing Agency that announced daily what the Standard Oil would pay for crude oil, and C. N. Payne. At Bradford on November 4 Sherman dined with General James B. Gordon, a member of the "Bourbon" triumvirate that dominated Georgia. Gorden informed him before lecturing that evening that Bryan's defeat for the presidency was attributable to " 'Government by Injunction' in platform & Tillman's sectional speech." [52] Sherman

[49] Sherman Papers, MS 1896 Diary, August 31, September 1, 16, 1896.
[50] Lloyd Papers (Winnetka), Sherman to Lloyd, October 3, 1896.
[51] *Ibid.,* M. L. Lockwood to Lloyd, July 27, 1898.
[52] Sherman Papers, MS 1896 Diary, October 2, 5, 8–10, 20, November 5–7, December 28, 1896.

had not approved of the nomination of the inexperienced Bryan and had been notably absent from the political rostrum in the Oil Region that autumn.

At Pittsburgh on November 19 Sherman attended a Pure Oil Company policy meeting at the "office of U S P L." First attention was given there by King, Jennings, Collins, Westgate, Lee, Phillips, Walz, and Liebel to the "Trust Contract" that he had drafted. Pending litigation was carefully reviewed. At the meeting it was "proposed to retain W. J. Bryan. James C. Carter of N.Y. preferred," Sherman noted in his diary.

Shortly afterward he was interviewed on the subject of the Trusts by M. Paul de Roussiers of the Musée Social of Paris, who was introduced to him by Lloyd. Sherman's well-defined views and voluminous information on that subject were of undoubted value to Roussiers in the preparation of his book. Then on the night of December 28 Sherman met with Emery and other leading independents at the former's Bradford residence to consider the Standard Oil's offer to purchase all the producing, refining, and pipeline companies in which Emery figured. "Price not satisfact. basis & must include all," they concluded as Sherman stated in his diary. He returned to Titusville the next day to confer with Lee and Walker to whom he "talked strongly . . . about perfecting Trust paper," an entry which alluded undoubtedly to his plans for the Pure Oil Company's transformation and the consolidation of the independent enterprises. While the law prohibited the sale of the main properties of limited-partnership associations Sherman believed that individual trustee contracts between them and the Pure Oil Company as a New Jersey corporation would be regarded as legal in Pennsylvania and elsewhere under the principle of interstate comity.[53] He reported to John Fertig that Emery and the United States Pipe Line Company did not consider it worthwhile to negotiate further "on the basis of price of S. O. Trust & Proposition must be open to all to accept." [54]

Interestingly, after Emery declined to sell out the independent

[53] *Ibid.*, December 28, 29, 1896.

[54] *Ibid.*, December 29, 1896.

movement to the Standard Oil, the price of Pennsylvania crude oil declined to below a dollar a barrel. By mid-June it was 87¢.[55] The lower price diminished the ability of the Pennsylvania independent producers to supply further capital for the expansion of the Pure Oil Company, and was the more remarkable since it occurred after the victory of McKinley and Hobart, which "Big Business" generally regarded as bullish. By that time the Pure Oil Company was expanding abroad so rapidly that within less than a year it had "houses & storage in . . . Hamburg, Berlin, Amsterdam, Rotterdam, Mannheim, France & England." [56]

Although Sherman's consolidation program thus met an unexpected obstacle, it survived and was carried into effect in 1900 by J. W. Lee and Clarence Walker, long his associates in the independent movement. The Pure Oil Company's capital was then increased to $10,000,000, and there was turned over to it in trust the majority of the stock of the United States Pipe Line Company, Producers Oil Company, Ltd., and Producers and Refiners Oil Company, Ltd. This achieved the integration of the independent business enterprises and the position of strength for them which Sherman had urged upon the Emery associates with Henry Demarest Lloyd's warm support for so long. The trust agreement vested the voting power of the majority shares of the three companies in fifteen persons for twenty years, while half of new shares subscribed had to be turned over to the Pure Oil Company trustees in the near future. The provision that three-fifths of the trustees backed by 60% of the shares of stock outstanding could remove any single trustee prevented Standard Oil attempts to buy in and control.[57]

Sherman would have opposed the negotiation of 1902 with the Standard Oil for a "division of territory, especially in Europe, and a chance to live" in which Lee engaged. The negotiation culminated

[55] *Butler Citizen,* June 17, 1897.

[56] Lloyd Papers (Winnetka), Emery, Jr., to Lloyd, March 23, 1898.

[57] Hidy, *op. cit.,* p. 269, erroneously dated this in 1897. Sherman Papers, C. Walker to "Dear Sir," June 12, 1900, J. W. Lee to "Dear Sir," June 23, 1900. Lee at this time was President of the Pure Oil Company. Tarbell, *op. cit.,* II, 189.

in an informal pool and, as Ida M. Tarbell soon described in her *History of the Standard Oil,* left price leadership in the possession of the Standard Oil Company of New Jersey.[58]

The last months of Sherman's career were spent in arduous preparation, in collaboration with Emery for the comprehensive criminal antitrust suit against the Standard Oil executives. Through John Cunneen of Buffalo he instituted this under the Sherman Act in the Federal District Court for New York City in July 1897. He cited in the bill of complaint a long series of injuries and unfair competitive practices perpetrated by the Standard Oil Company of New Jersey and its subsidiaries and allies against the United States Pipe Line Company, Producers Oil Company, Ltd., and Producers and Refiners Oil Company, Ltd.[59] Late that month Sherman asked Emery to have prepared immediately an inclusive list of damages inflicted upon the United States Pipe Line Company by the railroad allies of the Standard Oil, "by publication of false reports and libels referring to the business of the Company and its financial condition, . . . by reason of bribery of railroad officials and pipeline officials, employees and directors," including the purchase of Poth's plant in Germany, "by unfair and illegal competition, such as underselling in the markets, and all trickery of that kind which has affected the business of the United States Pipe Line Company." Sherman wanted this immediately so as to "form the basis of a general *claim* for damages." [60] It would be useful also in the antitrust case.

Sherman worked hard all summer in preparation for that proceeding, consulting precedents and legal authorities, gathering evidence. Emery supported him loyally in this. Success would have crowned Sherman's professional career with such a decisive triumph that the eminent judicial appointment that his admirers anticipated for him could hardly have been delayed. Success would have accomplished for the American business system, almost at a single stroke,

[58] Lloyd Papers (Winnetka), Lloyd to I. M. Tarbell, May 6, 1902; Tarbell, *op. cit.,* II, 191, 227–228.

[59] Tarbell, *op. cit.,* II, 186.

[60] Sherman Papers, Sherman to Emery, Jr., July 3, 1897.

what years of federal prosecutions during the future administrations of Theodore Roosevelt and William Howard Taft would have accomplished had not the United States Supreme Court promulgated "the rule of reason" that S. C. T. Dodd was urging repeatedly upon the legal profession during the "Mauve Decade."

Sherman, to be sure, was the ideal attorney to make the attempt. His long professional experience, his early association with the Gould group in the Oil Region with the inside knowledge that that had given him of the methods whereby the Rockefellers were given their first great boost up the tree of monopoly, his leadership in the Commonwealth Suits, and his years of counsel of the Emery associates had equipped him with an unmatched competence. If more were needed it had been supplied him by his five years of service as a Standard Oil attorney. No other attorney on the side of the independent oil men possessed the comprehensive detailed knowledge that he had of the multifarious illegalities that had characterized the Standard Oil's long war against its competitors. No other attorney possessed the fighting spirit, astuteness, and capacity to combine attention to immediate problems with a grasp of far-reaching consequences and fundamental principles. Emery was also a fighter. As loyal friends and business associates he and Sherman prepared thoroughly that summer for decisive battle in the federal courts.

For nearly a year Sherman had been unwell. His increasing weight, much greater than was wise for his age, and lack of exercise because of the burden of professional duties, had led him to go to Dr. C. S. Hubbard of Bradford. Sherman retained him "to fix me up," he remarked in his diary.

He was greatly fatigued on September 17, when he and Emery entrained on the Erie for New York to complete preparation for the great antitrust suit proceedings. When they arrived at the Erie Railroad station in Jersey City the following morning Sherman remarked that he was unwell. They went to the Astor House. At breakfast there the pain in Sherman's shoulder was so terrible that he had to go to his room. Emery put a hot-water bag between his shoulders and summoned a physician. Until half past one o'clock in the afternoon Sherman "did business lying down," as Emery informed Lloyd later. Sherman grew rapidly worse. Although Emery stood over him, alongside the best medical talent New York City could supply, Roger Sherman died of angina at 10:30 A.M. Sunday, September 19, his death the result, at least in part, of years of overwork in behalf of the independent oil men.

Emery's health was gravely affected by this tragedy. He informed Lloyd in late March 1898: "The shock was too much for me. My health was broken ere this trying ordeal & I went down, & have not yet got back & do not know whether I ever will."[61] This explains why Emery was not selected to head the enlarged Pure Oil Company two years later. As fighting leaders he and Sherman had been unmatched in the Oil Regions for their courage, persistence, and vision since 1872. Sherman's sudden death and Emery's virtual retirement, and the passing of the leadership of the Pure Oil Company and its associated companies to younger men, explain why the antitrust suit they had launched in July 1897 against the Standard Oil was never brought to trial.[62]

[61] Lloyd Papers (State Historical Society of Wisconsin), Emery, Jr., to Lloyd, March 23, 1898.

[62] Tarbell, *op. cit.*, II, 186, says that after Sherman's death the independent companies "had no heart for the suit, but allowed it to lapse."

# *Bibliography*

## MANUSCRIPTS

Confederate States of America, Military Archives (National Archives).

Crawford County, Pennsylvania, Court of Common Pleas. Appearance Docket. Appointment Docket.

James A. Garfield Papers. Library of Congress.

Henry Demarest Lloyd Papers. State Historical Society of Wisconsin.

Henry Demarest Lloyd Papers. Winnetka, Illinois, residence.

Mrs. W. B. Roberts, Diary, 1878–1880, 1881, 1882–1885. Courtesy of James B. Stevenson, editor, *Titusville Morning Herald.*

Roger Sherman Papers. Courtesy of T. W. Phillips, Jr., and Yale University Library.

Ida M. Tarbell Papers. Allegheny College Library.

Titusville Public Library Archives.

## PUBLIC DOCUMENTS

Committee on Manufactures. *Investigation of Certain Trusts,* House of Representatives Report No. 3112 (Washington, 1889).

Commonwealth of Massachusetts. *General Railroad Laws* (Boston, 1878).

Gowen, Franklin. *Argument of Mr. Franklin B. Gowen in the matter of the Investigation of the Standard Oil Trust by the Committee on Manufactures of the House of Representatives* (Philadelphia [February 20], 1889.

Interstate Commerce Commission. *Third Annual Report* (Washington 1890).

Pennsylvania Archives, Fourth Series, *Papers of the Governors,* IX (Harrisburg, 1881).

Pithole City School Board. *First Annual District Report, July 1, 1867* (Pithole, 1867).

*Report of Proceedings before the Committee appointed by the Pennsylvania Legislature to Inquire into the Legal Relations of the Standard Oil Company to the State* (Harrisburg, 1883).

## NEWSPAPERS AND PERIODICALS

*American Citizen* (Titusville), 1885–1889.

*Boston Herald,* 1895.

*Butler* (Pennsylvania) *Citizen,* 1895–1897.

*Chicago Tribune,* 1879–1880.

*New York Evening Post,* 1894.

*New York Herald,* 1872.

*New York Tribune,* 1872.

Patterson, Elisha G. *Newspaper Scrapbook.* Courtesy of Mrs. Eleanor F. Schreck, Independence, Kansas.

*Petroleum Age,* I–VI (Bradford, 1881–1886).

Sherman, Roger. "The Gospel of Greed: The Standard Oil Trust," *Forum,* XIII (New York, July 1892), 613–614.

*Titusville Daily Courier,* 1870–1875.

*Titusville Morning Herald,* 1868–1897.

Titusville *Sunday World,* 1889–1896.

*Utica* (New York) *Globe,* 1892.

## LEGAL DOCUMENTS

Commonwealth of Pennsylvania *v.* Pennsylvania Railroad Company, *Printed Testimony* [Titusville, 1879].

*Pennsylvania State Reports,* LVII–CLXIV.

### *Paper Books* in the Roger Sherman Papers

*Appeal of James E. Brown in the Case of George R. Yarrow, Trustee v. The Pennsylvania Transportation Company et al. Supreme Court of Pennsylvania, Western District, No. 51, October and November Term, 1880* [Titusville, 1880].

*Appeal of James E. Brown in Frederick W. Ames, Trustee v. Pennsylvania Transportation Company et al. Supreme Court of Pennsylvania, Western District, No. 4, April term, 1878.*

*Appeal of the Pennsylvania Railroad Company in Logan, Emery and*

*Weaver v. Pennsylvania R. R. Co., Supreme Court of Pennsylvania, Eastern District, No. 123, January term, 1890.*

*Appeal from Pennsylvania Transportation Company v. Pittsburgh, Titusville and Buffalo Railroad Company, Supreme Court of Pennsylvania, Western District, No. 308, October and November term, 1881.*

*Argument for Defendant, Pennsylvania Transportation Company v. The Pittsburgh, Titusville and Buffalo Railway Company. Court of Common Pleas of Crawford County, Sitting in Equity, No. 79 September term, 1880.*

*John J. Carter, Appellee v. Producers and Refiners Oil Company, Limited, et al., Appellants. Supreme Court of Pennsylvania, Eastern District, No. 82, July term, 1894.*

*Hascal L. Taylor, John Satterfield and John Pitcairn, Jr. v. John D. Rockefeller, H. M. Flagler, et al.* [Titusville, 1880].

*Henry Harley and William Warmcastle v. Pennsylvania Transportation Company, Appeal of Defendants. Supreme Court of Pennsylvania, Western District, No. 311, October and November term, 1880.*

## CONTEMPORARY BOOKS AND PAMPHLETS

*Address and Constitution of the Petroleum Producers Association* (Oil City, 1872).

*An Appeal to the Executive of Pennsylvania. An Address to Gov. John F. Hartranft, invoking the aid of the State against the unlawful acts of corporations, presented August 15, 1878* (Titusville, 1878.)

Bates, Samuel P. *Our County and Its People: A Historical and Memorial Record of Crawford County Pennsylvania* (Meadville, 1899).

*Catalogue of the Titusville Library Association* (Titusville, 1880).

*Charter, Constitution and By-Laws of the Titusville Citizen Corps of Titusville, Crawford County, Pa.* (Titusville, 1873).

"Crocus" (Leonard, Charles C.). *The History of Pithole* (Pithole, 1867).

*The Derrick's Handbook of Petroleum* (Oil City, 1898).

Henry, J. T. *The Early and Later History of Petroleum, with Authentic Facts in Regard to its Development in Western Pennsylvania* (Philadelphia, 1873).

*History of Butler County* (1895).

*History of the Rise and Fall of the South Improvement Company* (Titusville, 1872).

Jordan, Thomas, and Pryor, J. P. *The Campaigns of Lieut.-Gen. N. B. Forrest, and of Forrest's Cavalry* (New Orleans, n.d.).

Lloyd, Henry Demarest. *Wealth Against Commonwealth* (New York, 1894).

Pennsylvania Transportation Company, *Extension of Pipe Line to Tide Water* (n.p., 1876).

Sherman, Roger. *The Shermans. A Sketch of Family History and Genealogical Record, 1570–1890* (Titusville, 1890).

[Sherman, Roger, *et al.*] *History of the Organization, Purposes and Transactions of the General Council of the Petroleum Producers' Association, and the Suits and Prosecutions Instituted by It from 1878 to 1880* (Titusville, 1880).

Titusville Gas and Water Company. *Rules and Regulations* (revised, n.p., n.d.).

*The Titusville, Oil City, Franklin, Warren, Bradford, Olean, &c. Directory for 1878–1879* (Titusville, 1878).

## INTERVIEWS

With Mrs. Lena Emery Brenneman, Bradford, June 22, 1942.

With Miss Lucy Grumbine, Titusville, February 7, 1958.

## SECONDARY BOOKS AND ARTICLES

Benson, Lee. *Merchants—Farmers—and Railroads: Railroad Regulation and New York Politics, 1850–1887* (Cambridge, 1955).

Buente, Francis M. *Autobiography of an Oil Company* (New York, 1923).

Cochran, Thomas C. *Railroad Leaders 1845–1890* (Cambridge, 1953).

Destler, Chester McArthur. *Henry Demarest Lloyd and the Empire of Reform* (Philadelphia, 1963).

——. "The Standard Oil, Child of the Erie Ring, 1868–1872, Six Contracts and a Letter," *Mississippi Valley Historical Review,* XXXIII (June, 1946, March 1947), 89–114, 621–628.

Giddens, Paul. *The Beginnings of the Oil Industry* (Harrisburg, 1941).

——. *Early Days of Oil* (Princeton, 1948).

——. *Pennsylvania Petroleum, 1750–1872* (Harrisburg, 1947).

Grodinsky, Julius. *Jay Gould* (Philadelphia, 1957).

Helfman, Harold M. "Twenty-Nine Hectic Days: Public Opinion and the Oil War of 1872," *Pennsylvania History,* XVII (April 1950), 121–138.

Hidy, Ralph W., and Muriel E. *Pioneering in Big Business, 1882–1911* (New York, 1955).

Jessup, Walter. *Elihu Root* (two volumes, New York, 1938).

Johnson, Athur Menzies. *The Development of American Petroleum Pipelines. A Study in Private Enterprise and Public Policy, 1863–1906* (Ithaca, 1956).

Maybee, Roland Harper. *Railroad Competition and the Oil Trade, 1855–1873* (Mount Pleasant, 1940).

Nevins, Allan. *The Emergence of Modern America, 1865–1878* (*A History of American Life,* VIII) (New York, 1928).

——. *A Study in Power: John D. Rockefeller, Industrialist and Philanthropist* (two volumes, New York, 1953).

Schlegel, Marvin W. *Ruler of the Reading: The Life of Franklin B. Gowen* (Harrisburg, 1947).

Stocking, George Ward. *The Oil Industry and the Competitive System. A Study in Waste* (Boston, 1925).

Tarbell, Ida M. *History of the Standard Oil Company* (two volumes, New York, 1904).

Thorelli, Hans B. *The Federal Antitrust Policy: Origination of an American Tradition* (Baltimore, 1955).

Turner, Arlin. *George W. Cable* (Durham, 1956).

Warren, John E. "Release from Bull Pen—Andersonville," *Atlantic,* CCII (November 1958), 132.

Williamson, Harold F., and Daum, Arnold R. *The American Petroleum Industry, 1859–1899* (Evanston, 1959).

# Index